CALVES IN THE CLASSROOM

By the same author

Pigs in the Playground

Calves in the Classroom

by
JOHN TERRY

illustrated by
HENRY BREWIS

Farming Press Ltd
Wharfedale Road
Ipswich

First published 1987

Text copyright © John Terry 1987
Illustration copyright © Henry Brewis 1987

British Library Cataloguing in Publication Data

Terry, John, *1952–*
    Calves in the classroom.
    1. Rural conditions—Study and teaching
    (Secondary)—England
    I. Title
    630'.7'1242   S535.G7

    ISBN 0-85236-165-3

Typeset by Galleon Photosetting, Ipswich
Reproduced, printed and bound in Great Britain by
Hazell Watson & Viney Limited,
Member of the BPCC Group,
Aylesbury, Bucks.

CALVES IN THE CLASSROOM

A Farming Education

A STRANGE feeling came over me when I spotted the advertisement for the Head of the Rural Studies Department at my old school. I had started there at the age of four-and-a-half, and left when I was eighteen. I seemed to have spent most of my life at that place, and here it was, crossing my path again when I was looking for employment after completing my college course.

I sent for an application form, filled it out in my very best handwriting, and was delighted to be called for an interview.

I dug out my proverbial best suit, donned my white shirt and a 'tasteful' tie, and off I went. The candidates were seated in the school entrance hall, waiting to be called in. There were six of us – all plainly as nervous as anything. It was not quite so bad for me, I suppose, for I obviously knew the school . . . and the headmaster.

I was the last to be called in, after trying, without a great deal of success, to make intelligent conversation with some of the other candidates. It soon became apparent that the interview was to be a formal affair. The headmaster sat on the far side of his extremely large desk, looking at me over his glasses.

He proceeded to give me the 'third degree' – with a barrage of questions on the syllabus, my opinions on various teaching techniques, and other tricky matters.

1

Three-quarters of an hour later I was totally drained. He had really hammered me, and I had had to waffle my way out of some of the more wily questions. As I emerged from the ordeal I could not help thinking back to other interviews I had had to date. My farm 'interviews', for example. They were – shall we say – a lot less formal. . . .

It was 1970 and I was in my last term at school. After taking my examinations, I went for the obligatory talk with the careers advice lady, and was decidedly unimpressed. She looked about ninety, and the only career she had any real interest in, it seemed, was her own retirement. When I told her I wanted a job on a farm it floored her completely.

I explained that I would like to work for a year on a farm before going away to college. She appeared to be somewhat bemused by this, but glanced through her 'jobs on offer' book for something suitable for an oddball like me.

'I do have a job for a trainee plumber, young man. That's *manual*' – she said the word in what I felt was a disparaging tone. 'You'll be working with your hands. . . .'

At that point I gave up on the careers advisory service and literally 'got on my bike'. I cycled off to the nearest farm. I knew Mr Ronald Jones, the farmer, very well, and outlined what I had in mind. He said he had no vacancies, unfortunately, and suggested I go and see Bert, at the next farm.

I knew Bert, too. He milked cows, kept pigs, and grew wheat and barley. He had no vacancies, either – but thought that Mick Hardy, a nearby farmer, did. I thanked him for his help and cycled off to the Hardy farm.

I had never met Mick, but he had quite a reputation in the area for being a rather rough-and-ready character. He was noted, in particular, for letting the thistles grow in his grassland, much to the annoyance of his neighbours – like Bert – since the thistle seeds blew all over their land too.

Mrs Hardy answered the door and went back inside to fetch him. 'Hello, young man,' he said. 'You're lucky to catch me in the house. I just popped in to collect a screwdriver.'

I later learned that this was one of his favourite 'lies'. He spent most of his time sitting in the house, but like most farmers, was keen to perpetuate the myth of the 36-hour working day.

'Look, I've been to Bert's. I'm looking for a job, and he tells me you might have some work.'

'Yes, I need some help right enough,' he said, smiling all over his face.

He took off his slippers, put on his wellingtons, and joined me in the farmyard. It was a typical old-fashioned yard, a rectangle with buildings on all four sides, the house itself making up one side. At one corner stood a gate. Two of the sides were bordered by loose-boxes, and on the fourth side stood a rickety old cowshed. Outside that was a large manure heap.

Obviously the cowshed was shovelled out each day, and the manure heap, being just outside the door, was very convenient, even if a little unsightly. There were cockerels, hens, geese and ducks all over the place. They made a terrible mess – and an even worse noise!

'What jobs can you do, then?' asked Mick.

I explained that I had been interested in farming all my life, and had helped on Lord Clifton's farm. I could look after livestock, drive tractors and milk cows.

'Hah,' said Mick. 'Sounds promising.'

I suddenly felt quite elated, considering my lack of success at the other two farms. Not that I was under any delusions. The Hardy farm was not exactly in the vanguard of modern farming. But it was no more than a mile and a half from home – and I only needed a job for a year.

'I'll show you around,' said Mick.

We started with the cowshed. It was a long, low building with standings for only twenty cows – the sum total of his milking herd. The loose-boxes contained calves and pigs. I noted that the manure in the calf pens was a good four feet deep – he had nailed some extra boards on top of the doors to stop the calves jumping out into the yard. An interesting exercise in lateral thinking . . . but why on earth did he not just clean out the pens?

Beyond the farmyard was a rickyard. This consisted of a dutch barn, bulging with hay, half a dozen straw ricks and a larger building housing a few beef cattle. There was a small tractor shed, which contained no shining new Massey-Ferguson or John Deere, but a 'little grey Fergie'. It looked about a hundred years old.

Among the long grass and nettles I managed to glimpse a few farm implements. There was an old two-furrow plough, a rusty set of trailed disc harrows, a seed drill – which looked like the original Jethro Tull model – an old horse-drawn mowing machine, which had been fitted with a tractor towing hitch, a very old hay rake, an elevator without an engine, and an old manure spreader. The manure spreader had, in fact, lost

its vital parts and had been converted into a small trailer, one side of which was tied on with baling string, the farmer's quick-fix friend.

Not for Mick, apparently, the gleaming new combine harvester, the forage harvester, reversible ploughs, chisel ploughs or balers. If you had been able to persuade some poor unsuspecting soul to take over all this 'equipment' as a job lot, it would not have raised enough to buy even a second-hand combine. It was not that anything was actually vintage or 'antique'. It was just knackered.

Mick's farm was what the textbooks euphemistically call a 'low-input system' farm. He spent nothing, he boasted, on 'fancy pedigree livestock'. Instead he waited patiently at markets and bought whatever he could get — cheaply. His stock were fed on grass, home-grown hay and barley, and a minimum of bought feed. He used very little fertiliser and even less herbicides and insecticides. It was all a matter of principle: he was just plain mean.

His machines, he declared, were good enough for him — though some of them were literally held together by baling string and rust, and not necessarily in that order. He owned his 130 acres, so had no rent to pay. The farm, in fact, was under-stocked, and could have supported

5

many more cows and sheep. But Mick preferred to take his time. His age did not come into it; he was about sixty now, but he had had the same amount of livestock when he was half that age.

'How did you get the farm?' I asked.

'Oh, I advertised,' he said with a straight face. '"Farmer seeks a farming young lady with a view to possible marriage. Please send photo of farm and tractors."'

He roared with laughter. 'When I was twenty, I couldn't keep my hands off the wife – so I sacked the hands and bought a tractor!'

Mind you, credit where credit is due, Mick was rather good at making hay. His barn was positively bulging with the stuff. By my calculation he had enough hay to last him for five years. He relied on contractors to bale the hay and to combine his wheat and barley. Contractors would have to be his main expense, I concluded. He spent very little on home comforts, clothes . . . or his wife.

The last building he showed me was the farm toilet, a small, square, low brick building. Like everything else, it was the old-fashioned sort, with a wooden seat and a bucket. No fancy flush toilets for Mick.

'This is the workers' toilet,' he said, pushing the door open.

'But there's no lock on the door,' I observed.

'No problem, mate,' he replied. 'Nobody's ever stolen the bucket!'

On our return from the grand tour, he offered me the job. He would pay me the agricultural rate, he said, and I should start the very next day. I would take home the princely sum of £9 10s 6d each week, with more for overtime. There would be no perks like free milk or vegetables, however.

I turned up at seven the next morning. Mick did not appear, though, till gone eight. I reckoned he was probably looking for another screwdriver. As it turned out, it was par for the course.

Milking was always the first job of the day. I enjoyed that. My first outing on the 'little grey Fergie', however, was a disaster. There were animals everywhere in that farmyard – but when I reversed the tractor and trailer out of it, there was one less cockerel. As soon as I heard the awesome squawk I slapped on the anchors, but it was too late. The cockerel was as flat as a pancake, and the kiss of life was out of the question!

There was nothing for it; I would just have to come clean to Mick.

'Er . . . sorry, Mick, but I've just . . . er . . . run over one of your cockerels,' I stammered. 'I'm willing to replace him, though.'

Mick smiled an evil smile. 'Get your trousers off then, lad, and see what you can do with them there hens!'

I soon settled down into a daily routine. I would be there at 7 am, bring in the twenty cows, and milk them. Then they would be turned back out into the field while I cleaned out the shed and washed the milking machines and milk cooler. Mick would arrive just as I was finishing the last few cows. He would make a point of creeping up on me and trying to make me jump – his timing was incredible!

My next task was looking after the calves. Then I would feed the beef cattle and the pigs. I cleaned out the pigs about three times a week, and used the rest of my time in general work like digging out ditches or tractor driving. In the evening, the cycle of milking would begin again.

Mick had an interesting dog, Jess. She was like her master, a real character. She had a silly habit of holding on to the cows' tails and being dragged along until they kicked her loose. Mick would rant and swear at the poor animal, but she doted on him.

That dog had a fantastic built-in time clock. Unerringly, at 4.30 pm, she would come and find us and bark until we got the message that it was time to bring the cows in. If we took no notice, she would just go off and round them up on her own!

I worked really hard that year, but my work went largely unrecognised. Mick was not one to hand out compliments. I was not even allowed into the farmhouse – just as if I was not house-trained. In contrast, a part-time labourer named Bill always seemed to be made welcome. He would be ushered into the house for tea and coffee, while I had to bring my own, in a flask, and sit in the tractor shed to drink it. Even when there was a foot of snow on the ground it was just the same; never once did Mick invite me in for a warm by the fire.

Mick must have been worth a fortune. He spent next to nothing and gave nothing away, not even praise. Still, the job was only for a year and I was learning quite a lot about farming. Since it was a small and labour-intensive operation I had the whole range of jobs to do. I worked on my own most of the time, and was obliged to become adept at just about everything. I was learning fast about cows, calves, beef animals, pigs and sheep.

I soon came to the conclusion that Mick was cut off from the twentieth century as far as farming was concerned. He never read farming books or magazines – not even *Farmers Weekly*. The 'great debate' of the 1970s about entry to the Common Market was right outside his frame of reference . . . though he would never admit it.

'What do you think, Mick – should we join the Common Market?' I asked him one day in an attempt to stimulate a bit of intelligent conversation.

'Well . . . I don't know,' he reflected. 'As long as it don't clash with Rugby Cattle Market, on a Monday, or Melton Mowbray Market, on a Tuesday, or Leicester, on a Wednesday – then I haven't got any objections.'

One snowy day in January, Mick and Bill, the part-time worker, were cutting up some dead ash trees with the circular saw. Mick was slicing the timber into logs for the fire, and Bill was handing the branches to him. I was cleaning out the beef cattle at the time, and missed the drama that ensued.

Mick could not have been concentrating, because he suddenly sliced off the end of the first finger on his right

hand. He bound up the wound in his dirty old handker-
chief and they jumped into Bill's ancient Morris Minor
for a swift trip to the hospital.

The doctor in casualty took one look at the injury and
said: 'It's lovely, Mr Hardy. A nice clean cut. I can sew
the end of your finger back on – no trouble. Have you got
it with you?'

'Oh, yes,' said Mick. 'I've got it here, safe.'

He began emptying out the contents of his trouser
pocket onto the table. There were sweets, sweet-papers,
loose change, crumpled pound notes, two or three nasty-
looking handkerchiefs, his trusty old penknife, and
numerous bits of straw and other bits and pieces.

'Now then, where is it?' mused Mick. Blowing dust,
dirt and straw away, he picked up a very mucky-looking
finger end.

'Here it is,' he said proudly.

The doctor seemed distinctly unimpressed. Bill
fainted.

'He's useless,' said Mick, pointing at the prostrate Bill with his finger stump. 'Give him a cup of tea and a fairy cake, and he'll be all right.'

He was a real 'hard' man, was Mick. I've got to give him that. The next time he chopped off a finger he did not even bother going to the hospital. He just cursed profusely and threw it to the geese. The gander ate it!

After months of hard slog – I worked on Christmas Day, Boxing Day and over Easter – my stint with Mick was drawing to an end. Evening work and weekends were classed as overtime, and to be paid for this I had to keep a time-sheet. I would keep a record throughout the week and give it to Mick on Monday. With luck I would be paid the following Thursday or Friday.

I broke the news that I had been accepted at college and would therefore be finishing with him. On the last Friday he told me to estimate the hours for the coming Saturday and Sunday, so that he could pay me for them there and then. There was just routine feeding and milking to do over the weekend, so I guessed my hours for Saturday as 7 am to 10.30 am, with the afternoon feeding from 4 to 6 pm. For the Sunday I put down 7 to 10.30 am only, because I would need to have the afternoon free to pack for college.

I did my two stints on the Saturday as planned, and turned up to work on Sunday morning. I wanted to get through as quickly as I could, for there were a pile of things I needed to sort out before starting college on the following day. I dashed up and down the cowshed with the buckets of feed, and actually ran between the buildings. By 10.15 it was all done.

'I've finished now, Mick,' I announced. 'I'll be off, then. . . .'

'Just a minute, me lad,' said Mick. 'You put down on your time-sheet that you would be working here till 10.30 am.'

'Yes,' I said. 'But the work is finished, and I've got a ton of packing to do at home.'

'Oh no,' he announced. 'I can't let you go, just like that. Fair's fair, you know.'

For a fleeting moment the thought flashed through my mind that he might have a leaving present to give me – or even a surprise party. But that expectation was quickly dismissed.

'I'm not paying you for doing nothing!' declared Mick. 'You can clean the pigs out for the last quarter of an hour.'

My wage packet stayed firmly in his back pocket and he glared at me, arms folded across his chest, in a belligerent fashion. If I wanted my money, I was going to have to clean out the pigs.

I would not have minded so much if it had not been for the fact that we only ever cleaned out the pigs on Mondays, Wednesdays and Fridays. In the past he had never wanted them done on a Sunday, because that would have meant paying me overtime. I had never lost a day's work through absence or illness, and I had always been punctual. Yet here he was, determined to squeeze the last ounce of work out of me.

I just shrugged my shoulders – and got stuck in.

I started at college the very next day. It was a Rural Studies specialist course at Worcester College of Higher Education. I worked hard, and I loved every minute of it.

On the first day of the Christmas holiday, Mick phoned.

12

'How about giving up college, and coming to work for me?' he asked.

'What! I couldn't possibly do that. I've only just started.'

'Well, come and do some part-time work this holiday, then,' he pleaded.

'What will you pay me?' My curiosity was aroused.

'I'll give you the basic rate . . . plus an extra £10 a week . . . plus overtime, of course.'

'In that case I'll give it a try,' I said. 'I'll start tomorrow.'

'Great!' said Mick.

My parents were not exactly overjoyed at the news of my holiday job. In fact, they thought I was mad. I had told them how he had treated me, and they just could not figure out what possessed me to want to go back for more punishment. They were concerned, too, that I might neglect my college vacation work.

I managed, though, and my return to the farm brought an unexpected bonus. On my first day back, Mick was so pleased to see me that he even asked me, in a polite way, about college life. Then at break time I really knew that I had made it. He invited me into the house for a cup of coffee – the ultimate accolade!

I crossed those sacred portals for the very first time, took my coffee, and sat by the fire.

'Would you like a piece of toast'? asked Mick. I could not believe it.

'Er . . . yes, please.'

Mick noticed that I was looking at a pair of stuffed foxes in a corner of the room.

'Do you like foxes?' he asked. 'I shot them this autumn. The taxidermist asked me if I wanted them mounting. "No," I said – "one standing up and one sitting down will do nicely!" '

He roared with laughter, spilling his coffee. He got a fresh cup from a disapproving Mrs Hardy. I was amazed. Here I was, in the holy of holies, sipping coffee, eating toast, and chatting away. Things had certainly changed. . . .

But change of one sort or another was to be an essential part of my life over the next few years. After I obtained the post as Head of the Rural Studies Department came the adjustment to a schoolteacher's far-from-humdrum existence, with a whole new set of routines to be learned. The basic pattern now was made up of assemblies, marking registers, teaching lessons and attending staff meetings. It was all very formal in comparison with what I had been used to in those months on the Hardy farm.

As I sit in the staff-room today I sometimes look around me and wonder how some of our team of 'townies' would have got on with old Mick. I do not think many of them would have lasted long . . . or the headmaster, for that matter.

And I sometimes wonder how much I learned,

perhaps unconsciously, from Mick. He really was tight-fisted. I know I have a reputation for being 'careful' with money. But I could not be as tight as Mick . . . could I?

Follow That!

THERE it stood, the odd spot in the middle of an industrial town in North Warwickshire. My tiny agricultural oasis and kingdom.

Officially it was the Rural Studies Department in a comprehensive school of 950 pupils. To me, a frustrated farmer who had become a schoolmaster, it was truly God's little acre. But to the great big world of farming outside it was an anachronism – a quaint venture that had, incredibly, achieved a fleeting measure of national glory.

It was July 1982. I had just returned from the Royal Show with a trailer-load of champion sheep and a bunch of elated pupils. But a sense of anticlimax was already setting in. I gazed at our little 'patch' and pondered its past and its possible future.

Where, I wondered, do we go from here?

This school was a very special place for me. I had been a pupil here. Then after a skirmish as a farm worker I had returned as a teacher of Rural Studies – the study of the countryside and man's relation to it.

When I took over as departmental head, Rural Studies had been scornfully nicknamed the 'digging department'. It was a place where less able pupils and 'difficult' pupils might be kept out of mischief.

Tucked well away from the main buildings, it had

consisted of an acre of weary wasteland and a temporary wooden classroom. Well, over the years my pupils and I had certainly done a fair amount of digging. . . .

Gradually we had transformed the jungle into a smart, compact, registered smallholding. The stinging nettles, twitch and docks were gone. In their place we had created three grass paddocks and an orchard yielding apples, pears and plums. We were growing raspberries, gooseberries, blackcurrants and strawberries. We also had a well-stocked vegetable garden, lawns, flowerbeds, a rockery, a fishpond and a greenhouse.

Over eight years, by various means and much scrounging of materials, we had embellished our holding with several structures. These included a large agricultural building of 51 feet by 24 feet, and a brick building – erected in two stages – which now measured 25 feet by 16 feet.

We had built up a fine flock of pedigree Kerry Hill sheep with sixteen breeding ewes. Some of their lambs were being sold for the freezer and others as breeding stock on other farms.

The best we kept for showing – an activity we all enjoyed. We were now noted for our handsome sheep and we were renting a couple of fields to support our growing flock.

We also kept pedigree goats – two British Alpine females which we milked twice a day. Their milk had a ready sale to private customers, health-food shops and the two local hospitals, and their female kids were sold to other goat-keepers.

We were buying in half-a-dozen Hereford × Friesian calves at a week old, raising them to about five or six months and selling them to farmers to rear on for beef.

My pupils measured and weighed them each week and drew graphs of their growth-rate. A meticulous record of all costs was kept, and when they were sold the pupils worked out the profit – or loss.

Similar records were kept for the four weaner pigs we bought in at around six weeks old. These were sold at eighteen weeks to customers – including parents of pupils – who wanted them for the freezer. We were able to economise on their feeding costs through the generosity of local chemists' shops. They supplied us with out-of-date baby food on a regular basis. It was much appreciated by our pigs – in fact, they found the taste far superior to pigmeal!

The classroom recording process also applied to our flock of laying hens. Every egg was noted by the pupils and graphs were drawn to plot production through every month of the year. The eggs were sold to staff and parents, and here too, profits or losses were determined.

Our complement of livestock was made up by several ducks and rabbits – though the latter were kept purely as pets at the special request of the more sentimental pupils. The ducks were blessed with a pond which is probably the strongest ever constructed in Britain, having been established with a load of special concrete 'diverted' from a motorway scheme!

From all this you will have gathered that the department was plainly no longer the 'digging department' for backward pupils it had originally tended to be. All the second- and third-year pupils (confusingly, we have no first year at the school) were coming to it for one lesson a week. And it was an increasingly popular option for fourth- and fifth-year pupils. At the end of the fifth year they could take Rural Studies for 'O'-level GCE or CSE.

The subject now seemed destined to become even more popular, for we had returned from the Royal Show in a flush of fame after only our second season of showing sheep.

On the hallowed grass of Stoneleigh our ram had been awarded a third prize. That was pleasing enough . . . but our pair of shearling ewes had really swept the board!

They had won the first prize and also the cup as the show's champion Kerry Hill females, and that was not the end of it.

Our best ewe, Hazel, had then been teamed up with another exhibitor's champion male Kerry Hill. There were screams of joy from the Rural Studies contingent when they scooped the prize for reserve champion pair of any sheep breed in the whole show.

No wonder I was now feeling a little apprehensive about what we might do to follow that! As we unloaded our champions from the trailer, I thought back over the years that had led up to that achievement. Eight years of making do with second-hand equipment and materials. Eight years of scrounging and scheming. . . .

The department looked neat and tidy enough at first glance. But there was a lot to be done if we were to be really proud of it.

The trailer which conveyed our champions to and from the 'Royal' had not done so in champion style. The floor had started to rot. We needed a new trailer.

We also needed more permanent paths around the holding – muddy shoes were always making a mess in our classroom – and some new fencing. The Scots pine posts and half-round rails enclosing our three paddocks were going the same way as the trailer floor. Something more substantial was called for.

Then there was the old and overgrown snowberry hedge along the edge of the lawn. It looked awful! Perhaps we could dig it up and replace it with a large landscaped garden, planted with heathers and dwarf conifers?

I made a personal vow to try to improve my lessons – to make classes more interesting and to encourage my pupils to think more. But there were many other things in need of improvement which I could not achieve on my own. As I looked around the holding I found new problems at every turn. The wooden doors on our brick buildings, for instance – they desperately needed replacing. And inside, the walls had lost their whiteness and would have to be emulsioned. All the doors in the wooden building were crying out for a coat of paint, and the guttering needed attention.

And what about the routine jobs? Our apple trees would want pruning. The vegetable garden ought to be hoed soon – tomorrow if possible. That rose tree was most certainly dead and should be dug up. Those paving slabs needed to be re-laid, and the greenhouse plants would have to be sprayed again for whitefly. . . .

My thoughts turned to the stock and it was much the same story. The lambs could do with worming. The three goat kids were ready to be weaned and the pigs were ready to be sold. The hens' nest-boxes needed a good scrubbing. The duckpond would soon want cleaning out again. And oh – there was a rabbit who needed treatment for a cut foot immediately!

Somehow, I had to cope with all these problems *and* continue to show our sheep at about fifteen agricultural shows in the coming year.

Above all, I was determined to keep them at the top, and especially to get the female championship at the Royal Show again. The words of one of the exhibitors at the 'Royal' kept sneaking into my mind: 'It was just luck that a bunch of schoolkids won it.'

I intended to prove to him – and any other scoffers – that we were champion farmers who did not need to rely on luck to succeed!

The Sheep and the Police

HOWEVER, I did not anticipate that we would need police help in our attempt to prove ourselves champion farmers!

It had been a hard day at school, and I had dropped off to sleep the moment I hit the pillow. The telephone, I felt, had been ringing for some time before it penetrated.

I jumped out of bed on 'automatic pilot' and managed to grab the receiver after crashing into a table and staggering through the dimly lit debris I had dislodged.

The voice at the other end seemed to be coming through a mouthful of porridge: 'The skools seeps have gots out and vundered from the skool. The pleece have founds thems and brought thems back. . . .'

A practical joker from the school, I thought, until my mental switches started to click. Then I realised it was the school caretaker. He had either forgotten to put his teeth in – or had swallowed them.

'Pleece' . . . police! Funny how some words bring you up with a jolt.

'I'll be there in five minutes,' I said. I put on the light and took in the broken vase and water-soaked carpet. It was 1.30 am.

I felt a whale of a headache developing as I drove off into the cold, starless night. I was worried about Tessa and Tina, the new Kerry Hill ewes. I had last seen them at 10.30 pm before I went home to bed. Then they were grazing happily in the paddock nearest to my mobile classroom.

Kerry Hill ewes make perfect mothers, live long and produce lambs with a lean carcass. That was not my main aim in choosing the breed, however. The idea was to breed them to sell to farmers for breeding stock – and to keep the best ourselves for showing and breeding.

They are an attractive sheep, with sharply defined black markings on the face and legs. They have a black nose and a black patch around each eye, and black-and-white ears.

Our flock was now noted for its performance in the show ring . . . but what on earth had happened to these two new members? What was it the caretaker had said about the police finding them and bringing them back?

As I doubled the speed limit on my way to find the answers to these worrying questions, I reflected that sheep are prone to go 'walkabout' from time to time.

Mr Pryor, the farmer from whom I bought the two ewes, had told me that his flock of fifty had got out one evening – and trotted off to the nearby village! They had a really enjoyable munch at the local allotments before they were rounded up. The allotmenteers filled out claim forms stating what they had lost – eaten and damaged.

With a twinkle in his eye, Mr Pryor had reckoned that they must easily be the finest allotments in the country. It all totted up to ten thousand cabbage plants, a thousand onions and five hundred cucumbers . . . not bad going for fifty sheep.

I screeched to a halt in the school playground. In the blackness I could just make out a panda car – a Ford Escort – and two figures standing by it. The caretaker was talking to a policeman and he was kitted out in dressing-gown and slippers, and – sure enough – he had no teeth!

As I approached I became aware of two more figures outlined in the back window of the car. Crumbs, I thought, they are really playing this one for real; they have sent out the full squad.

Then it dawned on me that the two faces looking sheepishly at me from the back seat of the car were . . . Tessa and Tina. They were plainly two 'clients' that our man in blue could do without.

'I've had a terrible game with them,' said the policeman sternly. 'We had half-a-dozen phone calls at the station from folk reporting two sheep walking down the road. A motorist reported seeing them leaving the pub

car park. They sent me out, single-handed, and I chased them into a side street. It made quite a stir, I can tell you – people watching from their bedroom windows and me chasing the blighters all over the place. . . .'

I was beginning to think that he might even be enjoying the humour of the situation when he continued: 'A couple of drunks couldn't believe their eyes, and one lady, in curlers and a pink nightie, leaned out and shouted to me to make less noise.

'Finally I cornered them in a front garden – they'd stopped to sample the rose bushes. By this time a small crowd had gathered, and the only way out that I could see was to bundle the sheep into the car. Are they yours?'

'They certainly are,' I confessed.

'I felt such a fool,' he went on, agitatedly, and I could now see that there was nothing to joke about as far as he was concerned. I apologised as earnestly as I could and said I had left them secure and could not figure how they had got out.

We checked the paddock fence and discovered that one section had been vandalised; the posts and the linking rails were lying around in the field.

'They must have been big lads to break this lot down,' commented the constable.

'It can be fixed in the morning,' I said. 'In the meantime I have an empty shed to put them in for the night.'

I opened the police car door and lifted Tessa off the seat. Boy, she was heavy! Kerry ewes can weigh around 110 pounds – as much as I can manage, really. I was grunting with the effort of easing her out of the door and carrying her towards the shed.

The constable decided to hasten the operation by

picking up the second ewe. He promptly dropped her onto the playground.

'Look at this mess!' he yelled.

I put Tessa down alongside Tina, and went, in dread, to look. Nasty. . . .

The constable had thrown his hat into the back seat of the car when he went after the ewes. It was one of those flat hats – the sort with a chequered band.

It was still in perfect shape, not flattened in any way. Fine, really, apart from its contents – one inch of sheep's urine, still gently steaming and smelling pretty powerfully. Some of it had seeped through onto a plastic folder underneath.

At that moment even my shockproof watch was embarrassed. I secretly wished that the playground would open and swallow me up. The constable slid the folder from under the hat, gingerly, as if it might bite him. I doubt that he had ever changed a nappy. The liquid had soaked through many pages.

'These are official court documents and witnesses' statements,' he whispered. The ink had run and most of them appeared to be unreadable.

'It's a . . . disaster,' he said.

'Oh, I don't know,' I said comfortingly. 'The sheep could have made a far worse mess than that.'

The moment I said it I realised that I had put my foot in it again. The officer turned bright red, and the air turned blue! While making it very clear, in colourful and very unofficial language, exactly what he thought of me, he carefully lifted out his hat, as if it was a piece of delicate china, and tipped the contents into a nearby drain.

I wondered whether he could bring a charge against

me – or the sheep – for damaging police property. My mind conjured up a picture of Tessa, Tina and myself in the dock at the Old Bailey. Would I have to pay the court costs – or would the school pay?

The constable put his dripping hat on to the end of the car aerial to dry out, and placed the sodden documents in the front of the car near the heater. Then the three of us played 'sheepdogs' with the two ewes and, having them outnumbered, quickly got them into the shed.

'I'll give you ten out of ten for penning,' I said to the constable with a weak smile. Whoops again! He glared at me as if at an offensive foreign object.

'If you'd like to go and sit in my nice warm classroom, in my nice comfy chair, I'll be happy to clean out the car for you,' I said.

Our outside tap was out of action – the plumbing was being updated by the county council – but I reasoned that I could take the car to an outside tap at the other side of the school. That way I would get to drive a panda car. What a pity it was not one of those big 'jam sandwiches' – a Rover or even a Jaguar. I could put the blue flashing light on. . . .

My dream of joining 'Z Cars' was cut short by his curt reply.

'That won't be necessary. I've wasted too much time already. We have cleaners to do that sort of work, and I'll just drop off this car and pick up a clean one.'

He made the word 'clean' sound like an accusation, and the scene in the Old Bailey flashed back into my mind.

When he got back into the car he made great play of winding down the windows. It still smelt very 'sheepy', to say the least. They would certainly never sell it as aftershave. I just hoped that I would not be around when he switched the heater on.

He radioed the station to report that the ewes had been positively identified and returned to their owner. At the other end, it sounded as if a whole bunch of people were falling about. He then drove off without saying 'cheerio'. Funny, that.

I said goodnight to the caretaker and did the rounds once more. The calves were lying down, chewing the cud

– the commotion did not seem to have troubled them. The hens were roosting quietly on their perches. Only the ducks showed any sign of animation – they were awake and quacking. Sensitive creatures, ducks.

There was no further sign of damage. Pity about the fence, though. I paused and worked out in my mind how to fix it. I was glad to be back to some sort of reality. I still had a twinge of foreboding, however. Whatever would the headmaster say?

There was one cause for satisfaction, I mused. When I was a lad, I had kept and shown pure-bred rabbits, guinea pigs, bantams and cage birds, but I had always wanted to move on to bigger things – something more 'agricultural'. I would have loved some cattle to breed from and to show, but my total available land at school was just one acre, so I had had to be content with sheep.

Now I had a sudden vision of cattle charging around the town, and it brought on a strong feeling of relief to imagine what two heifers would have done to the constable's hat!

A few weeks later I came face to face with the constable when I was out shopping. I half wondered whether to make a run for it, but he had obviously spotted me, and was lengthening his stride and coming towards me through the crush of shoppers. The old heart moved into disco-rhythm again; I was in for a roasting, I felt sure.

'Apologies for the other evening,' he said.

'Oh!' I replied, somewhat taken aback. 'It should be me doing the apologising. Sorry about those sheep and the mess they made . . .'

'Well, to tell you the truth, it was just one of those nights,' he said.

'Sorry to hear that.' I could not believe my luck. We shook hands, and he started to grin.

'Do you realise, you've got the only two sheep in this area with criminal records? They're literally on the files, though I must confess that I didn't know quite what to put down on our villains' records.

'But it's nearly all forgotten,' he added philosophically. 'The only problem now is that every time I report for duty, some clown says "baa"!'

Gregory the Gander

My experience of keeping geese was almost nil, though I did have a little to do with the ones at the College. They were excellent guards – very noisy when disturbed. The gander was noted for his attacks. He would run very fast with his head lowered, at the same time making a terrible din. I did not take very much interest in the College geese because cattle and sheep were 'my cup of tea'.

My old farm boss, Mick, kept a few geese – they would wander around the farmyard and orchard. The grass provided them with enough food during the spring and summer, but in autumn and winter they were given poultry meal, wheat and barley.

On a bitterly cold day just before Christmas, Mick killed two geese, hung them up by their feet in an old cowshed and gave me the unpleasant task of plucking them. I had in the past plucked many fowl, but had only heard people say how difficult it was to pluck a goose and how time-consuming it was. This I was to learn for myself.

When I entered the old cowshed and saw the two white birds hanging there, I formed the opinion that Mick should not have killed them both at the same time. Poultry are easier to pluck when they are warm and as one goose would probably take me a long time the other one would be cold before I could make a start on it. But it was possible that Mick would give me a helping hand.

'I'll start you off,' he said. He pulled a few feathers out of a breast, then said, 'Now you have a go.' He could see I had pulled a few feathers out before and was not completely ignorant. But with a little concern he told me not to rip the skin as he wanted a good price for the birds and it was essential to have a good presentation. I smiled to myself. I would do my best to get the presentation right – but the conditions I was working in were far from hygienic.

The old cowshed was formerly used to house two dairy cows, who were tied up with chains around their necks. These were now rusty and covered with cobwebs and the concrete mangers were cracked and contained holes which led to underground mice nests. Before the cows had been housed there it had been a stable for a shire horse. The floor was cobbled and some of the stones had sunk deep into the ground with the weight of the horse over the years.

These days the old building was not used for anything in particular. Recently a couple of bullocks had been in it prior to going to market. I had mucked it out, but as there was only a single doorway the job had had to be done with a muck fork and wheelbarrow. Because of the unhygienic conditions, I decided to scrub the wall and floor where I would be working when Mick interrupted.

He said, 'It's a waste of time scrubbing the walls and swilling out, for when these geese are plucked there will be feathers everywhere.' Being only an employee I had to do what he said and that was to pick up the feathers afterwards and then clean the walls. I tried to explain my view concerning hygiene, but he was not interested. I am sure the environmental public health inspectors would have had a field day.

The building not only had filthy walls and floor, but when I gazed upwards, the rafters looked like a set from a horror movie. It was just as though the visual effects department had been busy, for the cobwebs up there were out of this world.

Mick then left me and, concluding that he did not intend to help any more, I set to work on the first goose. It was not very long before I had feathers in my hair and on my clothes, and an increasing pile at my feet. I found an old hessian sack, placed it on the manger and sat on it with the goose on my lap. My feet were so cold I could hardly feel them. It was really an enormous bird and its down – the small fluffy feathers – took ages to pluck.

Before commencing on the second goose I was so cold I had to jump up and down and run on the spot to get my circulation going again. I was plucking with alternate hands so that one could have a rest, but even so, I ended up with a sore thumb and index finger on both hands.

It took over three hours to do the two birds. The second took longer, it had started to go cold and was harder to pluck. I did not think I had made a bad job of them and the skins were not torn, but I did not receive a word of thanks or any encouragement.

The following morning I milked the cows and then decided to sweep up the feathers. It was bitterly cold and the wind was blowing. But if I shut the door I could not see what I was doing very well, so I had to have it open. With the wind blowing in I was again covered in down for it sticks to everything and gets everywhere.

I heard footsteps behind me and when Mick appeared in the doorway he remarked, 'Are you feeling "down" in the mouth?' He roared with laughter. I ignored him for he had done nothing to help me. The building was overrun with mice and I expect that during the night they had collected plenty of feathers to line their nests deep down behind the mangers. The plucking of Mick's birds did not increase my liking for geese, for it did not end there as I had to disembowel them and prepare them for the oven.

This coloured my view of geese so that later, when some of my pupils suggested that we should have some, my reply always was that we had enough animals to look after already.

Then three of the fourth-year pupils informed me they would like to study geese for their CSE project and would be only too willing to feed them before school had started in the morning, when it had finished in the afternoon, and also at weekends. I mellowed a little at this, and told them to go and find out more about them. The following morning, Eleanor, Lorraine and Anthony came to see me at break with a book on poultry, opened to the chapter on geese. Before long they were telling me how they would look after them and I must admit they earned full marks from me for enthusiasm.

I am not in favour of 'impulse buying' – that is, buying livestock without really considering and planning. I always convey this attitude to my pupils and I stress upon them that before buying any animal or bird they should purchase a book or borrow one from the school library or even from the public library and read about the subject. Then, if at all possible, they should go along and see the stock before deciding – but always buy the equipment first.

This is essential of course when buying a puppy. I do not like to see them offered for sale in pet shop windows, because I feel sure many people buy one on impulse. When they arrive home they realise they are not ready for it, not having equipped themselves with even the basic equipment, such as a bed or basket, collar or lead, and more than likely they do not know what to feed it on, how much is needed or how often. Other members of the family may not be in agreement. Arguments occur and possibly they cannot train it, get fed up with feeding it and in the end abandon it – often in a roadside lay-by.

After reading about geese I made some local enquiries but could not find any for sale; even my farmer friends were unable to help me. I consulted the classified advertisements in two or three farming magazines –

under 'Poultry' – sub-section 'Geese'. I telephoned a large reputable firm and explained my requirements, one gander and two geese to keep at the school. I stressed that I did not want them for killing at Christmas, but as a breeding trio for educational purposes.

After a lot of smooth talking, which included my 'hearts and flowers routine', it was still clear that the price was too high for what I was prepared to pay and that I was unable to obtain any discount. I think the manager must have had the same financial advisor as myself, for he gave me a taste of my own medicine. It was back to square one.

I then saw an advertisement from a farmer who was selling day-old Embden goslings, but unfortunately his address was in a small East Anglian village. It was certainly too far for me to arrange collection, so I hoped he could suggest delivery.

I thought the Embden breed would be ideal for us – they are a large breed with white feathers and orange legs and feet. They are recommended as a good meaty table bird – growing quickly and producing a high-quality carcase. They do not lay many eggs, which I suppose is a disadvantage, but I assumed they would lay enough eggs for our requirements.

Hopefully, as well as hatching the eggs naturally under the geese, we would be able to hatch some in our incubator and then sell them for breeding or for the table, or alternatively sell the eggs for eating. I thought some of the pupils would buy and try them – they are a very unusual egg to eat – they are enormous compared with a hen's egg and it is a certainty you would not want two of them on your plate for breakfast. Also the flavour is very strong in comparison.

A Mr Wilkins answered my telephone call, and after some discussion we agreed a very satisfactory price for three goslings – a gander and two geese.

I emptied a cardboard box that had contained packets of out-of-date baby food, kindly given to us to feed our pigs, and made a circular pen. There were two reasons for this – it would keep out any draughts and ensure the birds were not attracted to any corner.

I explained to the pupils that young birds will very often crowd into corners and possibly suffocate if there are sufficient numbers of them in the pen. This would be very unlikely to happen, admittedly, with only three goslings.

Our search in one of the cupboards produced our infra-red lamp. This was tested and found to be working properly. A further search unearthed a spare bulb.

We hung the lamp up at a height I thought would be about right. Goslings are hardy birds – they only need heat from a lamp for about two or three weeks. Approximately 30°C is ideal at first and the temperature is reduced to around 20°C after a couple of weeks. If the heat from the lamp is too fierce the birds will move to the edge of the pen. In contrast, if they are cold they will gather immediately under the lamp to try to catch the maximum amount of warmth – this means the lamp needs to be lowered.

We placed straw on the floor of the pen and found a shallow container suitable for water. We were unable to find a similar container for food, so we improvised with an empty egg box.

'With a bit of luck we could train them to lay their eggs straight into the boxes,' joked Eleanor. 'A good idea,' I replied. 'But have you seen the size of a goose egg?' Anthony chipped in, 'You wouldn't fit *one* in the box, let alone half a dozen.'

I bought some chick crumbs from our local pet shop on the Friday. We were now organised and eagerly awaiting the delivery early next week. The pupils came into school during the weekend and the subject of conversation was that they were all looking forward very much for Tuesday to arrive.

Mr Wilkins telephoned on the Tuesday morning. By chance I was in the school office so I was able to speak to him direct. 'I've just taken three goslings to Norwich Station and they will be leaving on the 9.40 am train to Birmingham, calling at Peterborough, Leicester and Hinckley on the way and they should be with you at 1.53 pm.'

'Thank you very much indeed,' I replied. 'The cheque was posted during the weekend, and if you have not received it by now I hope it will arrive in the morning.' Before terminating our call he told me he had marked the gander with a spot of paint on his head so that we could tell him apart from the other two.

The pupils were delighted when I told them the geese were on their way. The three pupils who had volunteered to look after the goslings were relative newcomers to the practical out-of-school work on the school farm. They had previously been coming down to the department before and after school, also at weekends – carrying out many jobs – but had not been in charge of any particular livestock. I was hoping the goslings would encourage them and give them some responsibility.

Just before the goslings were due to arrive we plugged in the infra-red lamp to get the area warm for them. The food and water were in their respective containers.

Arrangements were made with the deputy headmaster, Mr Bell, for me to be able to collect the goslings from the station – another member of the staff would take my class for the lesson. Eleanor, Lorraine and Anthony were also given permission to be absent from their lesson on condition they copied up the work they would miss. We all met at the staff car park and made our way in my car to the station.

The train was late and although we waited patiently on the platform, our time was limited for we had to return before the beginning of the next lesson. It would be very unprofessional of me to leave thirty pupils unattended.

As it happened the train was only seven minutes late. My pupils' faces were a picture as it pulled into the station. Various parcels were unloaded on the platform – including a bicycle – but there was no sign of the goslings. The happy faces disappeared.

'We are expecting a box containing three goslings,' I said to the porter.

'Sorry, there are no goslings on this train,' he replied.

'Oh, but there must be; a farmer by the name of Mr Wilkins telephoned me this morning to say they had left Norwich and would be arriving by this train. Can you have another look for me?'

'Come and have a look for yourself – but I know you will not find any.' I did, but there was not a sign of them.

I had to return to school, but before I left I asked the ticket collector when the next train was due from Leicester. He told me it should arrive at 2.38 pm. I managed to telephone Mr Wilkins before the start of my next lesson and explained the situation to him, but he could only reaffirm that he had taken them to Norwich

Station and they had been left there in ample time to catch the 9.40 am train.

He told me that he would make immediate enquiries at that end and ring me back. Before long I was speaking to him again and he told me that British Rail had assured him they had left on that train. His wife had filled in the labels which were attached to the box and he could offer no other explanation. I began to think they had been stolen, or even fallen off the train.

'All I can suggest is for you to meet the next train and see if they are on that one,' said Mr Wilkins. I explained the situation to Mr Bell, who agreed to take my lesson himself so I could pay another visit to the station. I conveyed the conversation I had with Mr Wilkins to the pupils, but as they had to go to their lessons, I made the journey on my own. The 2.38 pm train was on time, but there was nothing at all for us. I was again terribly disappointed. I returned to school and decided to telephone Norwich Station and the intermediate stations myself to find out what had happened to the birds. The parcel clerk at Norwich confirmed they had been placed on the train, and when I telephoned Peterborough Station, all I could ascertain from the staff was that they were not there – in fact, they had not seen them.

The telephone calls to Leicester and Hinckley proved negative, and taking into account the length of time involved for them to go and check to see if anything was there I was certainly running up the telephone bill.

Mr Bell was again very helpful and obliging, enabling me to make a further journey to the station to meet the 3.29 pm train. Luckily my three pupils were able to be excused from their lessons and so the four of us were eagerly waiting for the train to arrive. Again there was nothing at all for us amongst the parcels and mail bags.

'What shall we do now, Sir?' asked Anthony. 'I'll telephone the Hinckley, Leicester and Peterborough stations again,' was my reply. I did that on our return to school and although I could not get any satisfaction I had a feeling the British Rail staff were getting fed up with me, having heard my story before.

The next train due was at 4.26 pm and as the lessons had finished we were all free to make yet another journey. I explained to the ticket collector that we had already made three visits and bought platform tickets, and in view of this he condescended to allow us on to the platform without further payment. Patiently we waited for the train.

'Do you think we will be lucky this time?' asked Eleanor.

'I doubt it,' said the pessimistic Lorraine.

The train eventually arrived a couple of minutes behind schedule and we eagerly watched the parcels being unloaded. Success at last! Our box containing the goslings was placed on the platform. My pupils were delighted and so was I.

I signed for them in the parcels office and the clerk remarked, 'They have arrived at last.'

'Yes,' I replied, 'but it is a mystery where they have been until now.'

Lorraine held up the box to eye level and we all tried to peer through the small air holes. We could hear them making sweet little noises but could see nothing.

Lorraine carried the box. 'Don't drop it,' warned Anthony. 'Be very careful Lorraine, we do not want to lose them after all this aggravation,' added Eleanor.

We put the box on the back seat of the car and headed

for school. It was an exciting moment for all of us when it was opened and the goslings carefully lifted out. We put them into their pen, and they appeared to be none the worse after their long journey.

Eleanor moistened some chick crumbs and offered them on her hand to one of the females. She started to peck at Eleanor's finger, but when she found she could not eat that she turned to the crumbs. The other female was the next to be fed, followed by the gander. He seemed a little dopey, but this could be due to his new surroundings and the stress of the journey.

The next morning my pupils were waiting for me to

arrive, which showed they were enthusiastic. The goslings were fine and the pupils asked me what names they could give them. I told them it was their decision and after a brief consultation they decided the gander would be called Gregory and the two geese, April and May.

All three settled in very well and, after three weeks under the infra-red lamp, we put them outside during the daytime. They enjoyed eating the grass and although this would be sufficient to rear them we fed them with poultry meal, wheat and barley. As the weeks passed they lost their baby down feathers and these were replaced by their white adult feathers.

With the extra food they were getting they grew into large birds by the time Christmas came. A few of my pupils joked to the three looking after the geese that Gregory and his two girl-friends would make a good price per pound for the table. The reply they received was that it was hoped Gregory would father some goslings next spring.

Gregory was becoming a real character. When he spotted anyone, he would make a lot of noise, lower his head and neck and hiss. After Christmas, with the breeding season approaching, he was more daring. He would put his head down and run towards anyone in the vicinity, making a loud noise and flapping his wings. This usually frightened the pupils and they would run away, making sure they kept a good distance between them and Gregory.

He was only defending his territory and keeping guard on his two girl-friends, but unfortunately there was an incident when he did manage to hold onto a pupil's leg. When I was told of this I immediately fenced the geese in and ensured that only the three 'goose-keepers' and myself dealt with them. I explained to them that they had to walk towards Gregory full of confidence, and never retreat.

'When he puts his head down and starts making a noise, continue to walk towards him, wave your arms, shout at him and you will find he will retreat. Never take a backward step, otherwise he will think he is winning and will attack you.' It was easy for me to smile when pupils did run away because he did not frighten me – in fact I could pick him up and hold him in my arms. I suppose I looked a bit like Rod Hull and his Emu.

'Can we have a look around the farm, Mr Terry?' was a daily question.

'Yes, a pleasure,' I would reply.

'Oh – but Sir, is Gregory loose?' they would ask. 'I'm not looking around if he is on the rampage – he's horrible, Sir – a killer bird.'

Now I had erected a fence I was confident there would be no problems with the geese. To be honest I could not afford any – I did not want parents or governors complaining, because the geese made excellent guards for the school.

Then one Tuesday afternoon I went to the school office to order some food by telephone, and on the way back to my classroom I heard the geese. The noise they were making did not come from the school farm, but from one of the other entrances at the top of the school's back drive. As soon as I stepped onto the playground I could see that the geese had cornered Miss Warner, the needlework teacher, at the top of the two steps outside the two glass doors. The geese were hissing and making a terrible noise.

She appeared to be trapped. I could not understand

why she didn't open the door – maybe she was panicking too much and all she could think of was defending herself. She was doing this with her basket, which was full of exercise books.

I ran towards the geese but before I could get to them Gregory made a sudden grab for the basket, almost pulling it from Miss Warner's grasp. He was flapping his wings and she was very upset and frightened. I shouted at the geese who immediately turned away and made a hasty retreat.

'Mr Terry – get rid of them immediately – they will

end up killing me,' she said. After a few moments to compose herself she bellowed at me, 'I want danger money to work at this school. I was walking across the playground and heading for this entrance in order to visit the school library, when suddenly your geese ran towards me, especially that big brute – what do you call him – George is it?'

'Gregory,' I said.

'I would have made it – but would you believe it when I got to the doors they were locked and I couldn't open them.'

'I'm very sorry indeed – it was a good job I was passing,' I said.

'No – it's a bad job you let these vicious creatures out in the first place. Why don't you keep them in a pen or in a fenced-off area?'

'They were in a fenced-off area, which I thought was escape-proof,' I hastened to reply.

'Well I don't think they dug a tunnel to get out,' she said sarcastically.

'I'll check it out,' I said.

'Yes, please – they could have given someone a heart attack.'

I apologised once again and assured her they would not be seen on the playground again. By this time the geese had walked back to the school farm. As I ran back to have a look at their enclosure, Miss Warner shouted, 'How are you going to do that – roast them?'

When I examined the pen I found that the gate had not been shut properly, allowing the geese to escape. I fixed a new bolt and lock and made it more secure.

I am pleased to say the two geese laid more eggs than we expected. Some were hatched in our incubator and the goose we had named April built a nest and sat on nine eggs for thirty days. Eight of them hatched, making Gregory a proud father.

The lovely goslings were soon favourites with the pupils. They looked so sweet and innocent at that age. I could not help wondering if one would turn out to be another Gregory.

A New Trailer Is Needed

WE NEED to use a trailer most weeks of the year. In fact, my car thinks it is its birthday when it is not towing one with some load or other. My car suffers, and so does my pocket – unfortunately the county council will not pay all my petrol expenses for the hundred and one trips it has to make on behalf of our school farm.

And now I was girding myself for a really big buy – another trailer. Our original vehicle was well past its best. It had done a lot of hard work travelling miles and miles – often in a desperately overloaded state – behind my poor old car.

We needed one in tip-top condition, for travelling to shows, fetching weaner pigs from a farm each term and later taking them, as porkers, to the abattoir; for collecting and delivering our calves, going to sheep auctions, transporting hay and straw, rolled oats and barley . . . and a lot more besides.

At first we had managed without a trailer. I soon found out that this was a mistake when I urgently needed transport to get our first two pigs to the abattoir.

One of my friends, Stan, helped me out. We borrowed his wife's mini-van for the job, with dire results. The van caught fire on the way to the abattoir and we almost had roast pork before we even got to our destination!

Stan's wife was displeased, to say the least, so when I

saw an advertisement for a secondhand trailer in the *Farmers Weekly* I was very keen to take a look at it. To do this I had to travel all the way to Blackpool, and my trip to that famous place of fun was the nearest I had been to a holiday in years.

It was an aluminium eight-foot-long, two-beast, single-axle trailer, with a wooden floor, and it was in fairly good condition. We paid £375 for it, towed it home, and over the next two weeks painted the framework and the back with two coats of Land-Rover blue.

It looked smart and we were very pleased with it. There was one disadvantage, however. Because it was a single-axle model, much of the weight of a load at the front of the trailer was transferred to the rear end of the car. There were some anxious moments when calves, sheep or pigs moved forward in the trailer.

But the main reason why my car would never be the same again was our sales of manure. I invariably overloaded the trailer with bags of muck – one big load instead of two normal loads meant that I could finish more quickly.

Such a load, however, meant that the car's exhaust would almost be touching the road – and the front would stick up in the air to such an extent that I could hardly see over the bonnet.

I lived in constant fear that I would get a puncture, perhaps late at night coming from a show with eight or nine sheep aboard.

And now I had yet another problem. The trailer floor was gradually going rotten. I was increasingly conscious that it might not be safe.

I had made my mind up. I needed a new trailer, and it must be aluminium, with an aluminium floor. And it must have twin axles. Then it would still tow on three wheels if it suffered a puncture and its load would stay put and not play havoc with my car.

After all, our farm was now doing well. I had made do and mended, borrowed and bought secondhand tackle for years. It was time for a new status symbol!

I phoned three well-known manufacturers and discovered that the one I wanted would cost £1,125 after discount. Our farm fund contained the princely sum of £100. We obviously needed a lot more, in addition to the proceeds from the sale of the old trailer.

I decided to start with a plea to the county council, and after three tries I actually succeeded in speaking to the man at the top.

'It's Mr Terry here, head of the Rural Studies Department,' I chirruped. There was a moan and a sigh at the other end of the line. He knew me well.

'We're working really hard, and saving up,' I said. 'But we haven't got enough money for a new trailer, and we desperately need one. . . .' I went on and on, and the poor chap could not get a word in at first.

When he did it was 'No'. Undaunted, I soldiered on with my well-rehearsed 'hearts and flowers' routine. In the end he melted slightly. 'I'll do my best for you,' he promised. 'But I couldn't justify a thousand pounds or more on a new trailer. However, I will hopefully be able to allocate something towards it.'

I thanked him, but I did not expect too much. My guess was that we would get £50 or £100 . . . perhaps.

There was nothing else for it – we would have to go it alone with a fairly extensive fund-raising campaign. I soon came to the conclusion that oven-ready chickens would get it under way.

A New Trailer Is Needed

We had already had some useful, and profitable, experience with chickens – buying in plucked birds, dressing them, and selling them to parents and staff. It was now the right time of year, September, to do it in a bigger way. There was a new intake of 240 second-year pupils, all keen to obtain house points. House points had done the trick for us before, when we needed to extend our large wooden agricultural building. . . .

At the start of four of the second-year lessons, I announced that we would be selling oven-ready chickens on September 27 and October 4, and beseeched the pupils to try to get orders for those dates. The other four groups of pupils were told to seek orders for collection on October 11 and October 18. They all seemed keen on the idea of buying a new trailer. They seemed even keener on my offer of one house point for every two chickens sold.

I was privately aware that the headmaster was not at all likely to agree to this deal, as house points are normally given for good classwork and homework. Still, he would not find out, with luck!

For the larger orders I offered free delivery, and as it turned out my best seller certainly needed hers delivered – she took orders for fifty-four birds. The idea was working again, only this time better than ever; those chicken orders were rolling in! I was soon crowing over the lengthening lists which contained each pupil's name and the number and weights of the chickens they wanted.

I phoned Crawford's poultry farm, giving them a couple of weeks' notice and explaining what I was up to. Soon afterwards I received a welcome visitor – David, one of my former pupils.

51

David had been one of my 'rough-and-ready' pupils who had achieved some notoriety – and a severe wigging for me – by chasing a pig into the school home economics room. The pig joined a team of school inspectors who were having tea there, and made a pungent mess on the floor. That episode nearly changed my career . . . the headmaster threatened me with the sack!

David's unpolished demeanour was coupled with a heart of gold. He was out of work with nothing much to do, and he jumped at the chance to help with the chickens on two of the days – without pay. He was as keen as I was to get the new trailer.

The chicken project was now going so well that I began to have tremors about whether we would be able to fulfil the orders on time. The big problem was the second date – October 4 – when we would have an impossible task to prepare the 300 birds on order. I decided to do half of them on the previous day, and David agreed to come in on both days.

Into battle we went. The first week, with only eighty birds to do, was fairly easy going. Then we faced the 'big' week, with 300 to do in two consecutive days.

I collected 150 birds at 8 am on Thursday, October 3. David and half a dozen of my best 'gutters' turned up bright and early and our marathon began.

I missed assembly again, hoping that the headmaster had not noticed. He hated any member of staff missing assembly. To miss one in order to disembowel chickens would most certainly be a terrible crime in his book!

My best chicken 'gutters' had somehow managed to persuade their teachers to let them miss lessons so they could help. Their hurried letters asking for time off looked like blood-spattered ransom notes.

I had another team of pupils washing and weighing the birds, writing the labels and filling the bags. We worked away like demons. Break times were ignored.

I soon filled our refrigerator – and the refrigerators in the home economics department and in the staff-room. The latter was half-full of packed lunches. I moved sandwiches, tomatoes, apples and cartons of yogurt into a corner, ruthlessly – after all, I had to get my priorities right. I dared not allow our chickens to go 'off'.

I hoped to put more birds into the science department's refrigerator, but found that it was being used to store some weird bacterial cultures, and I did not fancy mixing chickens with them!

I had now run out of refrigerator space and the production line was still churning them out. I piled them into boxes which I then stacked in the coolest and most hygienic building I could find – the home economics room.

We did not stop for lunch and miraculously, by 3.30 in the afternoon, every bird was in its bag. We gathered them all into the classroom and arranged them on the desks, in the alphabetical order of the customers' names.

It looked like a poultry market and at a quarter to four it turned into bedlam as more than a hundred pupils rushed in and began a mad scramble to find their orders.

I sat by the door, shattered, as they argued about whose chicken belonged to whom. I wanted to go home, but this was the crucial period; I could not let anyone out until their name had been ticked off. Some stopped to pay me, but most would pay after the weekend.

'Your pound is worth more at Terry's store!' shouted one excited pupil carrying four chickens under each arm.

'It's the Great Chicken Bonanza!' cried another.

I could not hear myself think, let alone speak – it was like a riot at a cattle market. But within twenty minutes the crowd had vanished. I was left with three chickens. Only three pupils had forgotten to pick up their birds.

The next day brought a repeat performance with another 150 chickens – except that we were exhausted from the outset. I did not do much teaching in those two days; pupils arriving for classes were left to work out of books, or were lured onto the production line.

Luckily for me the headmaster came nowhere near, and had no idea what was going on. I could easily imagine his opening line if he had: 'What would the county council think – you are paid to teach, not run a butcher's shop!'

We worked hard over those four weeks. The other

sessions were not quite as hectic as those two peak days, but the result was spectacular. We sold a total of 835 oven-ready chickens – far more, according to our supplier, Mr Crawford, than the throughput of two large local chain stores in the same period.

In fact, their sales must have suffered as a result of our competition.

Anyway I was happy, because we had made an excellent profit. And a few days later I was jubilant when the school secretary gave me a letter from the county council. I had been granted £600 towards a new trailer.

I began to ask around my farmer friends to see if anyone wanted a secondhand trailer.

'Mr Finney's looking for one – try him,' advised the chairman of the Young Farmers' Club.

On my way to Finney's farm I recalled how we had upset him a few years back when our new ram, James Bond, got into his field of ewes and served two of them. When they both produced unmistakable little 007s, Mr Finney was not pleased. As they grew into handsome specimens, however, his grumbles abated.

He was working on a tractor that would not start. 'What are you on the scrounge for this time?' was his idea of a friendly greeting.

'Me . . . scrounge? Never let it be said! No, I hear you're looking for a good secondhand, two-beast trailer. I'm selling the school's trailer.'

'What's wrong with it, then?'

'It's a good one. It's just that I do about three thousand miles a year to agricultural shows, and I really can't afford the slightest risk of a breakdown. It would not look good if I was stranded in mid-Wales with a couple of my fifth-year girls! I need to sell it and get a new one.'

'Fair enough,' he said cautiously. 'But I know you, John; always very sharp, trying to make a lot of money. I do want a trailer . . . but I expect you would "do" me.'

As if I would. . . .

'To be very honest,' I said, 'it does need a new floor. But apart from that it's excellent.'

'Come on, John – what else is wrong?'

'I've told you. Nothing. Come and see for yourself. How about Saturday morning?'

On Saturday morning he got straight down to business. He viewed the trailer from front and back, got in and jumped up and down to test the floor, checked the wheels, crawled under to inspect the chassis and electrics, and finally examined the drawbar.

'How much money do you want for it?' he asked.

'It's worth £320,' I said confidently. 'That's very fair – a new one would cost you £1,125.'

'I'll give you £275.'

I looked hurt and shook my head. Deep down I was thinking that £275 would probably do. It was surely better than advertising it and enduring the hassle of people calling to view at odd hours, many of them merely time-wasters. And I knew I would have no problem getting the money from him.

'I need the £320 to put with the money I've saved,' I fibbed.

'It wants a new floor,' he pronounced, adding, reluctantly, 'but I might be able to manage £285. . . .'

I started to walk away.

'Three hundred,' he said. 'That's a nice round figure.'

I turned back. That would do, I thought . . . but I had

to try again. 'Three hundred and fifteen,' I said firmly.

'You're a bloody hard businessman, you are! I've already gone up £25, but you, John . . . oh . . . you've only budged a bloody fiver!'

It was time for my little sorrowful voice. 'It's just that I need to buy this new trailer,' I persisted. 'My pupils will be very disappointed if we don't get it.'

'Look, John. I'll give you £310. And that really is my top limit.'

'Done!' We shook hands on it.

I travelled to Birmingham to pick up the new trailer. It stood, all bright and shiny, in the dealer's yard . . . and it looked lovely. I gave Mr Nixon, the dealer, his cheque – it was the most we had ever spent on a single item.

There was just one thing wrong with the trailer. The ramp at the back had no springs fitted, which meant that it had to be lowered gently and lifted by hand – a heavy job for the pupils. Mr Nixon said that springs were an extra on the small models, and they would cost £18 for the pair.

I left it. We would have to get the springs when we could afford them – I had used up every penny of our cash to buy the trailer and a much-needed supply of feedstuffs.

The new trailer towed like a dream. Back at school I parked it alongside our old faithful. It looked good owning two of them. I was in two minds whether to move it before Mr Finney came to collect the old one – perhaps the contrast between the old and new might make him change his mind.

'This is the one, is it?' he quipped, pointing to the new vehicle.

'You can have it for fifteen hundred,' I replied cheekily.

He hitched up the old trailer to his van and gave me the cheque for £310. 'Don't forget you could do with a new floor,' I told him as he prepared to drive off.

'Don't worry about it,' he grinned.

At four-thirty that afternoon the school secretary came running across the playground, very red in the face and completely out of breath.

'There's a Mr Finney on the phone in the school office,' she spluttered. 'And please hurry because he's in a telephone box.'

'Thanks very much.'

'Mr Terry, I don't want to moan, but it's really time you found a farm secretary. Only this morning someone rang me to ask if you had any pigs to sell. And these sales reps are an absolute nuisance.'

Her lament was lost in the wind as I hared to the main school building. She had a point, I thought.

Jim Finney was obviously in a spot of bother. 'Can you come and help me out, please?' he said.

'What's the matter?'

'Well . . . I was moving a cow in the trailer to my other farm. Two miles along the A5 I heard a crash. It seemed as if the trailer brakes had come on. I stopped immediately and found that the cow had dropped through the floor – she was actually, standing on the road.'

'Is she injured?'

'No, John. She's all right. But I must have dragged her a little way along the road.'

'I'll be with you in ten minutes.'

I hitched up the new trailer, thinking it was a pity it

was going to get dirty on its first outing. I soon found Mr
Finney and his cow. He was peering into the back of our
old trailer.

'I know what you're going to say,' he said in a guilty
fashion. 'You're going to say that I could do with a new
floor, aren't you?'

I could not help using his parting line to me when he
took over the trailer. 'Don't worry about it,' I said.

I backed my trailer up to his and dropped both ramps.
The big Friesian, sorry for herself, stayed put.

'Come on, girl – move!' said Jim Finney. But she did
not fancy getting into another trailer and having the
bottom fall out of that one, too.

Mr Finney walked around her and shouted, and she

made a spectacular jump, off the road and into my new trailer, in which she promptly christened the floor with a steaming dollop.

We sorted out the broken floor and I followed him to his other farm, very pleased with the way the new trailer was working. We unloaded her – the cow kicked up her heels and disappeared down the end of the field.

'Thanks very much for helping me out, John,' said Mr Finney. 'But trust you to sell me a trailer with a rotten floor . . . it could have been very nasty, you know.' He said this knowing full well I had warned him!

My chance to acquire two springs for the ramp of the new trailer came at the local agricultural show. The dealers were there and I devised a plan to get hold of the springs from a trailer parked on their stand.

Four pupils were with me at the show – two husky fifth-form lads and two small and demure second-year girls. I told the boys I would not be needing them for half an hour and instructed the girls to follow me.

We walked up to the stand and said hello to Mr Nixon. 'Have you won any prizes with your sheep?' he asked.

'Yes,' I said. 'We've won the cup for the champion ram.'

'Well, you wouldn't have won that without our lovely new trailer, would you?'

'Certainly not,' I smiled.

'How are you getting on with it?'

'Excellent,' I said. 'However, there's something missing from it. It's like this: these two little girls can't get the back down – it's too heavy for them, and they stand no chance of lifting it up again. That's right, isn't it, girls?'

·'Yes, Mr Terry,' said Laura, responding to my wink. 'We really would like some springs on the back.'

Caroline nodded agreement. 'I hurt my back last week trying to lift it up!'

Mr Nixon explained, with some sympathy, 'Sorry, love, but I did say to Mr Terry when he bought the trailer that the springs are extra. Send me eighteen pounds plus postage and I'll send the springs – unfortunately I've no spares on the stand.'

'That sounds all right,' I said, 'but it's taken nearly all of our capital to buy the trailer. And I wouldn't have a clue how to fit them.'

Offhandedly, I added: 'I've noticed a pair of springs on that trailer over there. Could we buy those, please?'

'Oh, I couldn't do that, Mr Terry,' he replied. 'I hope to sell that trailer this afternoon.'

'Tell you what,' I said. 'How if I come back to see you just before you're ready to go. You might not have sold it, and perhaps you could sell me them then?'

'That's a good idea,' chorused my two girls – they were learning fast.

'You win,' said Mr Nixon. It was clear that he was only too pleased to get rid of us.

At four-thirty we were back at the stand, after telling the two lads to stay out of sight.

'Have you sold the trailer?' I asked Mr Nixon.

'No. . . .'

'Sell me the springs then, please,' I said. 'Oh, do sell us the springs!' said the girls in unison.

'If I sell you the springs it means I've got to fit new ones on when I get back,' he growled.

'Oh, go on,' I entreated. 'I don't want these little girls straining their backs and I'm sure you don't. . . .'

'I did hurt my back last week,' said Caroline, wide-eyed and reproachful.

'All right – you win!' said Mr Nixon glumly. 'I'll get some tools.'

Our victory seemed a little hollow when he appeared with the news that he had locked himself out of his car and it would probably be ages before his wife came back with her keys.

But I was not beaten, yet. 'I've got some spanners in my car,' I informed him.

'I want a hammer as well,' he said.

I tossed him my spanners and dashed to the next stand. 'Haven't got one – try that stand selling tractors.' All they had there, however, was a large wooden mallet, and that was no good.

I ran back to Mr Nixon to say: 'Don't go away. I'll be back as soon as I can.' He wiped his brow and shook his head as if to say, 'If ever a man suffered. . . .'

Three more stands, and still no hammer. Suddenly I had a bright idea. I zoomed to the opposite end of the showground. The steam engine enthusiasts had their display there, and the first one I asked thrust a hammer into my hand without waiting for an explanation.

I ran back to Mr Nixon as fast as I could, brandishing it triumphantly. He soon transferred the springs to our trailer.

'That will be eighteen pounds, thank you,' he said.

'Wait a minute,' I said. 'They are secondhand . . . I'll give you fifteen.'

'Wait a minute!' he snorted. 'They aren't really secondhand – that was a new trailer they were fitted on. And I've fitted them on yours. That will be three pounds, so we're back at eighteen.'

'Hold on,' I said. 'Don't forget that you used my spanners – and I've had to hire the hammer. Plus it's a cash sale.'

'You win,' he said. 'Fifteen it is . . . you can have them for fifteen pounds for your bloody cheek . . . oops, sorry girls, I didn't mean to swear.'

'It's all right,' said Caroline.

'We're happy now,' said Laura.

'I should think you are,' he sighed. 'I should bl . . . ooming well think you are!'

Harry the Hoover

THE SHEEP arriving at the school in the police car was a real-life drama but sheep continued to cause problems.

My marking session ended suddenly when Christopher came running into the classroom. 'Sir, Sir, I think Jenny is going to lamb at any minute!'

'I'll come and look straight away,' I said, putting down my red pen.

It was 4.15 – and a lambing at this time of day was very pleasing. I could give her my undivided attention; I had no classes to worry about, and I would not be burning the midnight oil.

Hawthorn Jenny 2nd, a home-bred two-year-old ewe, was lying down, pushing and straining. After a while she stood up and walked around in a circle, pawed the ground with one of her front feet, and lay down again with her nose pointed upwards. The pushing and straining began again.

'I'm right – she is going to lamb, isn't she?' asked the excited Christopher.

'Yes, she's going to lamb,' I replied.

We were soon joined by half a dozen other pupils, trusty helpers who often stayed on after school to help on our little farm. Now their chores were forgotten – a ewe about to lamb was far more interesting than cleaning out pigs and mucking out calves.

We all sat and watched quietly. This would be a good lesson for them.

The ewe's water bag had burst and I was waiting to see two little hooves poking out. Christopher had fetched a bucket of water and a box of Lux soapflakes.

'You'll probably need these,' he said.

'I certainly will – she doesn't appear to be making much progress,' I replied.

I took off my jumper, rolled up a shirt sleeve and lubricated my arm with soapy water. Christopher held her head and I moved her tail to reveal a tiny lamb's tail. There were 'oohs' and 'aahs' from my audience.

Inside the ewe I could feel the lamb's hindquarters. Its back legs were tucked right under its body in what is called the breech presentation. I explained this to the pupils and added that I would do my best. I could not help looking at their faces – they were now extremely tense and worried for the wellbeing of the ewe.

Very slowly I pushed the lamb back into her, making sure I pushed when she was not straining. I found a rear hock and pushed it up and backwards, then held a back foot and found it was easy to bring this forward. I carried out the same procedure with the other hock and foot.

I remembered the golden rule – that once the lamb's tail is out, you must rotate the lamb and pull it out quite quickly. At this stage the navel cord is stretched and held between the lamb's underside and the pelvis of the ewe – and if the oxygen supply to the lamb from the ewe is cut off, the lamb will die.

In no time I had delivered a fine ram lamb. He coughed and spluttered and began to breathe – success!

I quickly moved him to the front of the ewe. She had

not lambed before, but she took to him and began to lick him. I put some iodine on his navel to stop him getting joint-ill. He was a good size – I estimated he would tip the scales at about fourteen pounds when he was weighed for our records.

The pupils thought all this was marvellous. Most of them had not witnessed a lambing before, let alone a classic breech presentation. They had all been very sensible, and I, of course, was feeling very pleased with myself.

The lamb's markings were excellent and I already had visions of showing him. We moved him with his mother into one of our lambing pens – small pens which we had constructed from hurdles. This was the ideal place for them to get to know each other.

Jenny licked him dry and he was soon on his feet and, after falling over a few times, was looking for a drink of colostrum. I pointed him in the right direction and within seconds he grabbed a teat and was sucking hard and wagging his little tail.

It was a happy scene. Half an hour later the ewe began to pass out the afterbirth. She would not have another lamb, but I did not care – I was more than satisfied with this one.

Soon the lamb, after a brief lie down, was ravenous for another drink. He flopped down, like a little barrel, after filling his belly, and I began to fear that he might be heading for a stomach upset.

'He's a real boozer,' said Christopher.

'Yes – he's a greedy little devil,' I agreed – little realising, at this stage, how apt that description would turn out to be!

After going home for tea, I returned to school and checked the ewes again at 9 pm. There was no action yet with the others which were due to lamb, but with Jenny it was a different story. Suddenly she walked over to her newborn lamb and head-butted him. He fell to the floor, got up and walked towards her, looking for a drink. Bang! She butted him again, knocking him for six.

I could not believe it – only a few hours before she had licked him dry and let him drink; now she did not want him. She bowled him over yet again and I began to get worried that she might injure him.

I had visions of having to keep them apart with a wire grid, and having to hold her four or five times a day while the lamb filled his belly, until she took to him. With other recalcitrant ewes, I had successfully used the old trick of putting a dog in the pen with the lamb and ewe. In this situation the ewe regards the dog as an enemy who will harm the lamb, and will often protect it. I might need to borrow a dog tomorrow, I thought, but there was still another ploy I could attempt. . . .

I picked the lamb up – and suddenly she stamped her foot as if to say, 'Leave my lamb alone!' At this I made a noise like a crying lamb and moved away, making believe I was stealing her offspring.

This brought her bustling to the front of the hurdles; she was actually looking for him, and that was very good. I repeated my imitation of an upset lamb, and now she was baaing for him – she wanted him!

Even so, when I put him back in the pen I half-expected her to butt him again. Instead she began to mother him, licking him gently, and after a minute or two she let him have a drink. She was a new mother and very confused. Since then I have tried stealing lambs from other ewes which did not want their lambs but it did not work; they still would not take to them.

He was obviously delighted with the reunion, and he made the most of it – he did not stop drinking for ages. When he could not hold another drop, he waddled to the corner of the pen, lay down and immediately went to sleep.

After this episode the ewe behaved very well. She became an excellent mother.

Because of my knowledge of livestock, the pupils regard me as a sort of 'cheap vet', bringing all kinds of pets for my expert opinion. I have handed out diagnoses on mice, hamsters, guinea pigs and rabbits.

Some of the pupils even seek a consultation about their own sore throats, warts, spots and abscesses. A nice-looking lady member of staff asked me what she should do about her septic toe. I suggested that she should bathe it in hot salt water – and went on to recommend a wonderful new sheep injection which I felt was sure to cure it.

She went off the idea, however, when she saw the size of the syringe and needle, and I explained that the injection was intramuscular – and so I would have to inject her in the backside!

But back to our 'boozy' lamb. After a few days I placed him and his mother with a small group of ewes who had all produced lambs with show potential. My practice is to bring them indoors during the evenings, and to provide the lambs, day and night, with access to some creep-feed.

It became more and more obvious that Jenny's lamb was a very greedy lamb indeed. The pupils nicknamed him Harry the Hoover – he certainly cleaned up any food that was put in front of him. He was always munching away at the creep-feed – a feeder with bars set so close together that the ewes cannot get at the contents.

But Harry was not content with mother's milk and creep-feed. He soon got into the crafty habit of creeping up behind another ewe, shoving his head under her, and snaffling as much milk as possible from a teat before she could realise that he was a hi-jacker.

When she did, she would invariably butt him out of the way. But Harry was tough, and he could stand being butted – after all, he had suffered this in his younger days. He thought the extra milk was worth the aggravation, despite the fact that his mother was never short of it. It was simply that he wanted more . . . and more.

Naturally he grew very well indeed, and it became my aim to show him at the Royal Show in July. It took only a couple of weeks to halter-train him. He had always been remarkably tame – he would come running to you at the slightest rustle of a toffee-paper. He loved sweets, chocolates, sandwiches, crisps and apples, and defied his herbivorean heritage by even eating the ham from a sandwich.

We shampooed him a month before the Royal and we weaned him a week before the show – and hefty Harry made a dickens of a noise for half a day when he had to part from his mum. I do not think he was missing her so much as her milk supply – and that of the other ewes as well!

I carded and trimmed him half a dozen times, and just before we left for the show on the Sunday evening we washed his face and legs. This year we did not expect to win the glory we had enjoyed a couple of years before; however, we were hoping to get a rosette or two. Before unloading Harry and his three companions at the showground we placed hessian sacks inside the pens, so that our nice clean sheep would not rub against dusty, dirty or rusty metalwork.

I placed our older ram, James Bond, in the first pen, our shearling ewes Amy and Amanda in the next, and Harry in the end pen. He hated it! It was the first time he had been completely on his own and he placed his front feet on the bar of a hurdle and baaed and baaed so loudly you could hardly hear yourself speak.

'When did you wean him?' asked Mr Porter, a rival exhibitor.

'A week ago,' I said.

'You should have done it sooner, then he would not have made all this noise,' chided Mr Porter.

'I suppose I should have,' I said. But deep down I knew it would not have made much difference – Harry was spoilt and had a mind of his own.

Other exhibitors were raising their eyebrows, shrugging their shoulders, and trying to shout above the din.

'How long have we got to put up with this noise?' asked another exhibitor, Mr Williamson.

'I don't know,' I replied guiltily.

Harry was certainly letting us down. I had given him some food but amazingly he ignored it.

'I'm sure he'll settle down,' I said confidently to the pupils. Twenty minutes later he was still bleating like a strangulated foghorn. In desperation I put him in with the two shearling ewes, hoping he would settle down in their company.

What a hope! At once it became clear that Harry was determined to be the master of them – a sultan in a little harem. His top lip curled back and he launched into an attempt to serve each ewe in turn. They did not take kindly to the brusque advances of this inexperienced adolescent. They did not want him as a boyfriend, and they pushed him away as he tried, again and again, to mount them.

It was no good . . . Harry just would not give up the assault. I caught him and moved him in with his father, James Bond. It was my last chance to settle him down.

Harry immediately put his head between his dad's back legs and furiously rummaged for a teat and a drink of milk.

Mr Porter leaned over the hurdle. 'He won't find much milk there!' he roared.

James Bond acted with all the ruthlessness and speed of his namesake. He butted mightily, smashing Harry up the side of the hurdle. Harry shook himself and made another try for the milk. This time James Bond retaliated with another bone-cruncher, followed by two or three more for good measure.

Harry had now stopped making a noise – what with one thing and another, I suppose he had developed a splitting headache! He sniffed around the floor of the pen, somewhat dazedly. I gave him some hay, and he settled down.

I had to teach all day on the Monday, but I quickly loaded the car with pupils after school and headed for the Royal Show. We found a couple of old ladies standing at Harry's pen, feeding him crisps. He stood with his front legs on a hurdle, just like a goat. James Bond was lying down – he had no interest in old ladies and bags of crisps!

Next day was judging day and I had plenty of helpers. I had chosen Matthew to take Harry into the ring. He donned his white coat, pinned on his exhibitor's number, put Harry's halter on, and joined the circle moving around the ring. The lambs were lined up and the judge began his inspection. Harry was easily the largest lamb on parade – I bet he ate more than the rest put together!

The judge looked them over for size, and for markings – lambs with large black markings do better than brown- or grey-marked animals. He looked at the ears – Kerry Hill sheep should have upright ears, not floppy or droopy ones. He checked length and depth to the body, and inspected teeth.

The ram lambs were all led in a circle again; then they were called back into line, this time in order of merit. I was delighted when the judge called Matthew in to stand with Harry in third place. I was, of course, only a spectator, watching along with pupils and parents.

There was still a chance that the judge would change his mind. But no . . . he signalled to the steward and it was all over. Mr Porter's lamb had won and Mr Stone's was second. Our Harry was indeed third.

'Well done, Matthew,' I said. 'You handled him very well indeed.'

'I'm very satisfied,' said Matthew. 'But do you know, Harry was looking for food the whole time he was being judged – he even sniffed the judge's pockets for sweets!'

Our 'normal' sheep also did well. James Bond was fifth and Amy and Amanda were third. However, those awards did not mean quite as much to me as Harry's prize. Not only had we bred him, we had bred his mother, too. This was our first award with a sheep that was second-generation bred, a landmark for our tiny flock.

The prize did not change Harry one bit. He was still a glutton. He loved to find titbits in the field, such as a plastic bag with a few crumbs at the bottom. He would even lick the inside of an empty crisp packet. This was worrying; I was scared that he would swallow a plastic bag, resulting in a serious blockage. After every lunchtime and break I sent pupils out in search of such dangerous litter.

All went well until one October afternoon. Harry the Hoover was living with two other ram lambs; they spent the daytime in one of our paddocks, but were moved indoors for the night, as a security measure.

After school, Matthew burst into my classroom. 'Mr Terry, Harry looks awful!' he cried.

'He looks really ill,' added Philip, who had helped Matthew to bring the sheep indoors.

My heart sank and I hurried to the sheep building.

'He only ate about two mouthfuls of food,' said Philip as we got to Harry, lying in a corner of the shed.

'Most unlike him,' I said, now extremely worried.

Harry stood when he saw me, but his back was arched and I could tell he had stomach trouble. He kept putting his head on one side and looking at his rumen. Then he lay flat out on his side and kicked his legs in apparent pain.

'Will he die, Sir?' asked Matthew.

'We will do all we can for him,' I said. But I was really at a loss to know what to do. He seemed to be suffering from colic. His temperature was normal. I drenched him with our vet's patent cure for colic – an old-fashioned draught medicine.

I checked on his progress before leaving for home, and again an hour later. By bedtime, when I checked once more, he had come from death's door and was looking better.

Both Matthew and Philip also returned after tea to see how he was; it was good, that – it made me feel that my pupils really cared.

Next morning we were relieved to find that Harry had made a complete recovery. When we turned the rams into the paddock, his two companions wandered casually through the gateway. To our amazement, Harry whizzed through it.

He travelled at what seemed to be getting on for a hundred miles an hour to the far corner of the paddock, stopped dead, and put his head down.

He appeared to be eating something. The three of us walked over to him, but he had no time to look at us – he was too busy filling his mouth with fallen damsons.

The over-ripe fruit had fallen from an overhanging tree in a neighbour's back garden. If there is a world record for damson devouring, Harry was breaking it with ease. He sucked them up, just like a hoover – stones and all. Here was the cause of yesterday's stomach trouble. . . .

My two pupils heard the school bell and ran off for

their registration. I was late for my register because of the time it took me to collect all the fruits which had fallen on our side of the fence. I filled three trouser pockets and two jacket pockets, and carried the remainder in my hands. In the classroom I unloaded the damsons onto my desk.

'Have you been scrumping, Sir?' asked Jennifer, slyly.

I put the damsons into a plastic bag and into the refrigerator, and started to mark the register. As I sat there calling out the names I became aware of a wet feeling on my backside and leg. My pupils roared gleefully as I ruefully took two very squashed damsons from my back pocket.

I sent them off to assembly – I had decided to give it a miss yet again! – and returned to the paddock, intending to pick the damsons from the branches hanging over our land.

Harry was still there – he was on his hind legs, hunting for more fruit. When a gust of wind blew a few off, he immediately got down and hoovered them up.

With one eye on the branches and the other on the neighbour's house I picked the rest of the fruit on our side and hurried off before the householder, Mrs Hunter, spotted me. I suppose I should have given them to her, but I had other ideas.

I felt I had done a good job. I had saved Harry from another violent stomach ache and now I planned to make good use of those damsons.

Farther around the department I spotted another neighbour, Mrs Miller, hanging out her washing. A farmer's widow, she was an active member of the Women's Institute and a frequent winner in their competitions with her superb home cooking, bottling and jam making.

'It's a fine autumn morning,' I said.

'Lovely morning.'

'It's been a good year for fruit,' I remarked.

'Yes,' she said. 'I'd love to bottle a lot more this year, and make some jam. But everyone seems to have the same idea – I just can't get hold of enough fresh produce.'

'You're in luck,' I said. 'Our damsons have yielded quite well. I would be willing to sell you our surplus.'

'I'd love some,' she said.

I ran back into the classroom and quickly weighed the two plastic bags of damsons. She paid me there and then – the going rate plus an extra fifty pence.

'Can you sell me a few more?' she asked.

'I'll check our tree. I think there are still some we haven't picked.'

I waited until dusk. As I entered the paddock I noticed hundreds of damson stones which had obviously passed straight through Harry. There were no more damsons on the overhanging branches. I leaned cautiously over the fence and 'scrumped' a few more. Only a few – I didn't dare risk getting caught. Whatever would the headmaster say, or my pupils, for that matter?

But I urgently needed to raise some cash to buy some feedstuffs for the stock. When I finished the tree looked a bit odd. There was loads of fruit on the far side and none

on the branches near to our land. But at least Harry would be protected from wind-borne fruit. . . .

I went with my haul straight to Mrs Miller and she was very pleased. In all I took just over ten pounds in money from her that day. Magic!

As far as I knew the tree had not fruited before. I made a note to watch it next year and to avoid a repeat of Harry's alarming performance.

A week later Mrs Miller shouted from the bottom of her garden – 'I've got a present for you, love.'

She passed a jar of home-made damson jam over the fence. 'Have that with my compliments,' she said. 'You were very good to let me have those damsons.'

'Thank you very much indeed,' I said. 'I'll certainly enjoy some of this for my tea this evening.'

Suddenly I was smitten with pangs of conscience. Perhaps I should not have charged her for those damsons. Still, it was all in a good cause. And anyway, it was all the fault of Harry the Hoover!

Crime Doesn't Pay

WHEN I started teaching at the school we had more than our fair share of rogues among the pupils – and I seemed to teach all of them. One of the most colourful was Nicholas Galloway, the school's motoring fanatic. During his fourth year he 'nicked' a car in the Co-op car park and was caught as he tried to drive it out – damaging a dozen other vehicles in the process.

When he left school, his first job with a painter and decorator lasted three weeks. He did not like the painting, but he was delighted to be driving the firm's van – they did not know he was borrowing it during lunchtimes – and was sacked after he had taken it off on an extended joy-ride.

While on the dole he got into more trouble for unauthorised joy-rides, and developed into an excellent illegal driver. He was on probation when he came to see me.

I suggested that he should join a 'banger' club – that way he could drive as fast as he liked without taking a car on the road.

He did not favour that. Instead he fell in with three men who planned to rob a sub-post office in Birmingham, and jumped at the chance of being their 'getaway man'.

At three o'clock in the afternoon the trio burst into the post office, stockings over their faces, and demanded money at gunpoint. They ran out with the money to find Nicholas standing dumbly by the getaway car. Unbelievably, instead of sitting there with the engine ticking over, he had locked himself out of it!

His prize comment at the scene – 'Is anyone in the AA?' – caused much merriment in court.

He was just the same dimwit at school. He could never get anything right.

As the years have gone by, however, our 'catchment area' has changed. Nowadays most of our pupils are pleasantly well behaved; though we will, I suppose, always get the odd one or two 'problem' pupils.

One of these was Adrian Mills, a rogue throughout the four years he attended the school. He always seemed to be in trouble. He did not plan his nefarious escapades, but got up to mischief on impulse, and consequently he was usually caught.

He was not the only serious offender, by any means. Mark Reeves and Neil Webb were also in his category, to name just two. Adrian was probably the worst of the bunch, though. He had been in trouble with the police and I knew that I could not trust him and needed to watch him carefully.

One day, when my second-year class were all seated in my mobile classroom, listening to my opening remarks on geranium cuttings, the fire bell sounded.

It was sure to be a fire drill, I thought. The usual well-rehearsed procedure was for each teacher to accompany the group of pupils they are teaching, shepherding them quietly but quickly out of the nearest door leading into the playground.

Once there, the pupils leave their teaching group and line up in their respective forms – the second-years nearest the school drive, then the thirds, fourths and fifths. The form teacher checks the register to make sure that everyone who appeared at registration that morning or afternoon is still present. Usually, during a fire drill, this check is made very quickly – the bell by now having stopped ringing – and the 'all clear' is given for everyone to return to their classrooms.

This time, though, it was different. The bell continued to ring after the form registers had been checked, and the fire brigade and police arrived and went into the school.

'Is it a fire, Sir?' I was asked.

'I've no idea.' I was as puzzled as they were. There was no sign of smoke or flames anywhere.

Something *was* going on, however, and I learned what it was from Mr Petty, the woodwork teacher. The school secretary, he said, had received a telephone call from someone who said there was a bomb in the school!

It seemed that the fire brigade were taking it very seriously, too. One of the fire officers advised the headmaster, Mr Beech, that the pupils should be taken well out of the way, for if there was a bomb, and the oxy-acetylene bottles in the metalwork room exploded with it, there would not be much left of the school. . . .

Mr Beech forthwith instructed us all to leave the premises. Each form was marched, in an orderly fashion, to a building site opposite the school gates.

The pupils naturally revelled in the drama and excitement – and, of course, they were missing lessons. I would sooner have been teaching Rural Studies, suspecting, along with many others, that it was all a hoax. But I appreciated that with 950 pupils and fifty staff you cannot take any chances in situations like this; everything must be done by the book.

After about twenty minutes the headmaster blew a whistle, signalling that we were allowed to go back into school. The premises had been checked by the experts and the verdict was that it *was* a hoax.

After all the excitement my next class, a group of fifth-years, were not in a mood to settle down to serious work. I must admit that I was edgy, too. I glanced out of the window and watched a panda car being driven across the playground – with our arch-rogue, Adrian Mills, sitting in the back!

My class noticed him, and the comments were not slow in coming – 'I bet it was him, Sir'; 'He's the hoax caller, all right.'

When lessons ended I went into the staff-room to find a group of teachers in an unusual state of hilarity.

'What's so funny?' I asked Mr Petty.

'Ha-ha – well, Mrs Loveridge, the school secretary, had this telephone call that there was a bomb in the school.'

'Yes, I know that.'

'Well, when the police came she told them the call was at ten-past two – change-of-lesson time. It sounded like a young boy and it came from a telephone box, she said.

'They checked the telephone box opposite the school and there, lying on the shelf, was Adrian Mills's library card!'

'Talk about being booked!'

'Yes. When they interviewed him he admitted making the call. He's been taken down to the station.'

'I saw him in the back of the police car,' I said.

'Fancy leaving his name in the telephone box – of all the stupid things to do. He just can't get anything right!'

Adrian certainly did not seem to plan his 'jobs' very

professionally. It was not surprising that he was blamed
for another bit of malpractice that occurred on the very
last day of term.

Traditionally the headmaster then holds a final
assembly for the whole school, in the only place big
enough for it – the gymnasium. We finish lessons early
and, beginning at twenty to three, a series of bells ring to
summon tutors and their forms to the final assembly.

It was the end of the summer term and we were all
looking forward to the six-week break. When everyone
was in the gym, Mr Bell told the pupils to sit on the
floor. The staff stood round the sides of the hall. When
silence had been achieved, the headmaster mounted the
stage and began his final speech.

'It is traditional for me to stand here and remind you

84

of the successes the school has achieved this term,' he said. 'I am pleased to say that through various fund-raising events the school has collected and raised over a thousand pounds for charity.

'That is a real credit to the school. Now I want to mention our sporting achievements – they are second to none in the area.'

The headmaster went into detail regarding all our sporting wins, and then moved to the Rural Studies Department's achievements with sheep at various shows.

Then suddenly his expression changed – from cheerful pride to dark depression. For a split second I thought he was going to say that the Rural Studies Department had once again spoilt things by allowing livestock to escape and get the school into trouble.

'I'm sorry to have to inform you that I am going to end on a very sour note indeed,' he continued, with an expression signifying grievous pain.

'Most pupils at this school are good, hard-working, caring people and – as you have heard from our achievements – they make me proud to be headmaster of this school. However, there will always be one or two who spoil things for the rest.

'Today is no exception. The large iron gates at the end of the school drive have been stolen. They are very large gates indeed, each measuring six feet wide and seven feet high.

'They are also very heavy, and I feel sure, because of their weight, that more than one of you are involved.

'They were definitely in place this morning – Mr Bell noticed they were missing half-way through the lunch hour. Mr Bell and I have spent a great deal of time looking for them, but they are nowhere to be seen.'

Mr Beech now adopted his most stern mien – the one he reserved for situations of the utmost seriousness.

'Some pupils in this room have played a practical joke which is just – not – funny. Those gates would be most costly to replace. And what on earth will Warwickshire County Council think of me being in charge of a school which has had its gates stolen during the last day of term?'

There was an awful silence.

'If you are responsible,' said Mr Beech, 'put your hand up now – and then the rest of the school can go home. I'll warn you now . . . if no one puts up their hand, I'm afraid I must ask staff to accompany their forms back to form rooms, and you will all wait there until the culprits confess.'

The silence was unbroken – and no hands went up.

'If anyone saw anything, I don't expect you to put your hand up now . . . but have a quiet word with your form tutor when you get back to your room.'

Mr Beech's voice rose dramatically as he pronounced his final words on the situation: 'I must – and I will – get to the bottom of this today!'

Back in the form room I asked if anyone had any clues. There was no immediate response, then Annette said, with solemn conviction: 'You can bet Adrian Mills is behind this, Sir.'

'Yes, Sir – he should be behind bars,' agreed Julia.

'Have you any proof?' I demanded.

'No,' they replied despondently.

The rest of the form were no help, either. I began to imagine us all still sitting here until midnight – the headmaster, I felt, had never seemed so determined. All I could hope for was that other form tutors had achieved more success at this detective work. I was not in a hurry to go home, unlike most of the pupils and staff, but I did want to get on with some pressing tasks on the farm.

Twenty minutes or so later, the bell began to ring again to summon another assembly in the gym. Mr Beech mounted the rostrum for the second time that afternoon. His audience waited with eager anticipation. Had he tracked down the culprits?

'I am pleased to say that I have got to the bottom of the matter of the school gates,' he began. There was a great sigh of relief from all who had feared a long and uncertain wait.

'The gates were not stolen,' continued Mr Beech.

'A few minutes ago the school caretaker informed me that they were officially taken on a lorry by Warwickshire County Council workers. They have gone to the depot to be repaired!'

His stern face creased into an unfamiliar grin, and then he actually laughed outright. So did the entire school! His statement, and the way he delivered it, really brought the house down, as they say in show business. Pupils were rolling on the gym floor, clutching their stomachs. Some of the staff roared until tears rolled down their faces.

It was a fine, rousing end to the term, after all.

Adrian Mills had, of course, been interviewed by Mr Beech. But he was plainly 'not guilty' this time. We learned that Mr Beech had personally apologised to him for suspecting him of the 'crime'. I had to admire the headmaster. I think I would have taken the easy way out and would have merely told the school: 'The gates have been returned – you can now go home.'

The long summer holiday went quickly enough and the next term was launched without incident – until the night my telephone rang just as I was about to get into bed. I was very tired, and could have done without the call from Mrs Miller, the maker of the damson jam, whose garden was among those which backed on to the school farm's orchard.

'Mr Terry, the geese have just been making a great deal of noise,' she said quickly. 'I've just walked up my garden path and it looks as if someone has pushed their way through my hedge and into your orchard. My hedge has certainly been disturbed in one place – and as I said, the geese made a terrible din.'

'How long ago was this?' I asked.

'Only a few minutes.'

The geese were indeed at the bottom of her garden, running loose in the orchard. But they were usually quiet at night . . . unless someone disturbed them.

'I'll come and have a look right away,' I told her. I would not be getting my early night, but this needed to be checked out.

It was a fine, warmish evening. Instead of driving in through the school gates and across the playground to my classroom, I parked along the road outside and walked quickly to Mrs Miller's house.

She was a widow who lived alone. When I knocked she inspected me through the front window, then opened the door.

'I'm sure there's an intruder,' she exclaimed. 'Someone has definitely gone through my hedge and got over into the school, and it's the first time I've heard those geese at night.'

'I'll have a look,' I said.

'Oh no, Mr Terry,' she said, with a great show of concern. 'Let's call the police – that's the best thing.'

It was, I thought, a sensible suggestion. But something made me say 'No' to it.

'Don't do anything silly, dear,' said Mrs Miller.

'I won't,' I assured her. 'I'll just check, first. It's probably only someone taking a short-cut through the school grounds.'

'Well, I don't think you should risk it,' she insisted.

Walking up her garden path, I could see by the moonlight that someone had gone through the hedge. The geese were still there – and they started to hiss as soon as I appeared, but shut up after a few seconds when they realised who it was. I expected they were disappointed by my appearance at the scene – they usually associate me with feeding time.

I walked through the orchard and crept behind the mobile classroom to peer through a window. There was no one in sight. I made my way silently around the farm buildings – nothing there, seemingly. Then I heard the noise.

I had just walked past our small wooden shed, which was stacked with hay. Yes, there was someone in there.

I had to go back past the shed – there was no other way to get out – and I began to feel a bit cornered. I heard the noise again, a loud rustling, and my mind whirled with possible explanations. Perhaps it was just a courting couple in there, on the hay. But then – perhaps it could be some murderous individual who would jump out and strangle me, or stick a knife in my ribs. There had been a gruesome murder in the locality not long ago, and no one had been caught for it. Soon I was imagining a whole gang of thugs watching me through the cracks in the shed wall and waiting to pounce.

It was no good, though – I had to walk past the shed to get back to the playground. I tried to pluck up courage, and looked around for a suitable weapon. There was nothing like an old spade handle or an iron bar available – I kept the area too tidy for that.

I braced myself, clenched my fists, and wished I was built like Sylvester Stallone.

'Come out!' I growled. 'I know you are in there!'

There was no response, which merely served to escalate my fear. What had Mrs Miller said about calling the police?

'Come out – now!' I roared in desperation.

With that the door moved and something shot out at what seemed like a hundred miles an hour. It screamed as it brushed against me like a ricocheting bullet, and disappeared into the night.

It all happened so fast it took quite a few seconds before I realised that my hideous mass murderer was . . . a cat.

I had frightened it. But not half as much as it had frightened me.

Slowly I came out of deep shock and tried to pull my wits together. That had been a cat, but it had not been a cat that had flattened Mrs Miller's hedge. That was a fact. So what next?

I took a long look at the main school building, set away across the playground from my department. It all looked quiet. I guessed that if there had been an intruder in the main school, I would surely have disturbed him with my stentorian challenge to the cat – and the cat, too, had made noise enough to waken the dead!

I was walking back to Mrs Miller's garden when I heard the sound of smashing glass. I felt my blood run cold – for the second time in a few minutes.

Hiding behind my classroom, I peered around the corner, across the playground. I could see nothing unusual, but I sensed that the smashing noise had come from the metalwork room, or somewhere close to it.

I waited for a couple of minutes, and then thought I could see the light from a torch, actually in the school corridor. I was not too sure, because the outside lights are left on at night and reflect into the school.

But the light was moving now, without any doubt. I ran back through the school garden and cut across the paddock to the caretaker's house, with the intention of getting him to phone the police. There was no reply to my knock – just my luck. I ran back to Mrs Miller's house.

'What's happening, dear?' she asked excitedly, sensing my urgency.

'It's burglars – I want to telephone the police at once, please.'

I did not wait for an answer, but raced into the living-room – I knew the telephone was there because I had used it once to call out the vet. I suggested that the police officers should meet me by the gates at the school's rear entrance – next to the caretaker's house.

Then I thanked Mrs Miller for sounding the alarm.

'You're not going back there, are you, dear?' she fluttered.

'Oh, yes,' I said.

'Well, just tell the police what's going on and leave it all to them – this burglar might be dangerous, or even armed. You read so much about this sort of thing in the papers!'

'I expect it's just one or two children, messing about,' I said in an attempt to calm her fears. I remembered, then, that Adrian Mills had recently been caught breaking and entering a house, and I must admit I did think it might be him.

A police car arrived in an amazingly short time. I explained the situation to the two policemen who got out of it. I also told them that the caretaker was out, and I had not got a key to the school.

At the very moment when I was telling them this, Bill Clarke, the caretaker, pulled up in his car with his wife.

'You timed that well,' I said. 'There's a burglar in the school and we could do with a key to get in.'

'Do you have a key, sir?' asked one of the constables.

Bill dug into his pocket and brought out a bunch of keys. He picked out one of them and held it up. 'This will get you through that small rear door,' he said, pointing to the door. 'If you both go in there you can split up – there's a corridor to the left and one to the right.'

The constables ran up the drive and disappeared into the school. I stood by the gate with Bill, wondering what we would do if the intruder ran towards us. Would we tackle him – and if so who would make the first move, me or Bill? I did not fancy it being me!

Within minutes, someone opened a window and jumped out of the corridor onto the playground. I could see he was carrying something bulky under his arm, though I could not make out what it was.

He started running very fast across the playground and then down the drive – right towards Bill and me. My heart began to thump and the adrenalin was flowing in torrents.

Suddenly he saw us, turned to the right and clambered over the fence into one of our paddocks. I could clearly see, now, that he was carrying the school's video recorder – the plug and lead were dragging on the ground behind him.

The caretaker and I gave chase, without exchanging a word. One of the policemen jumped out of a corridor window and soared across the playground, taking the fence into the paddock like a hurdler. He was closing fast with the intruder, the other policeman not far behind.

The intruder climbed over the next fence into our orchard – he was obviously going to make his getaway through Mrs Miller's garden.

He had only made a few strides across the orchard when the geese began to make a tremendous row. They were very annoyed at being disturbed again, and determined to defend their territory.

One of them rose up out of the grass, wings flapping like mad, and headed straight for the strange runner. It was Gregory, our malevolent gander.

In a flash, he had grabbed one of the intruder's legs – obviously the fleshy part behind the knee, for the man cried out in pain, stopped and kicked at him wildly.

One kick knocked Gregory away, but only for a fraction of a second. Soon we could see that Gregory had got hold of him again – and he was really hurting the fellow this time.

94

I do not suppose he could have held on for very long, but it was enough for the first constable to close in and make a flying tackle, rugby-style. As they both went down in a muddy patch, Gregory was still hanging on.

His wings were flapping and his two wives, April and May, were making so much noise you could hardly hear yourself speak. Bill and I and the other constable moved in to help, marvelling at the scene.

I fully expected Gregory to let go and attack one of the policemen – let us face it, he was not quite intelligent enough to be able to tell the goodies from the baddies – and I did not want to have a battered policeman on my hands.

I grabbed the gander and literally had to wrench him away from the intruder, who was now suffering from a bent arm up the back, and Gregory's bruising beak.

I checked Gregory and he seemed to have been unharmed by the kick – he was a huge, tough bird and I doubt whether he had even noticed it. I shut him in a pen with the other two geese, and he fluffed up his feathers, made a sort of 'well, that's over' noise and settled down to scoffing some wheat. He was fine.

In the meantime, when the policemen were leading the culprit away by his cricked arm he kicked out at one of them and tried to make a run for it. The handcuffs were quickly slapped on his wrists and the hurdling constable – he was not even out of breath – intoned: 'I am arresting you for burglary – you are not obliged to say anything, but if you do, it may be used in evidence against you . . .'

'What's that bird – that thing – that bloody monster?' demanded the villain, glaring towards Gregory's pen.

'Oh, that's only Gregory,' I said. 'He's a lovely bird.'

'I want to stop and have a look at my leg – I'm sure I'm bleeding,' he howled.

'I don't expect he's ripped your skin, but you'll probably have a couple of bruises the size of saucers,' I said.

'Serves you bloody well right, too,' said Bill. He had picked up the video and was wiping the mud off it.

'What were you going to do with the video?' one of the policemen asked.

'Sell it,' he moaned. 'I need the money – I've got a wife and three kids to support.'

We all walked back through the school garden, across the playground and down the drive. At the police car the second policeman reported the incident on his radio. The prisoner – whom I now recognised as a past pupil of the school – was bundled into the back of the car. The policemen thanked us for our assistance and drove off to the station.

Bill said cheerio and took the video into his home. He was anxious to tell his wife the news. I went back to Mrs Miller's and she listened, enthralled with the whole story.

Next morning Bill reported the incident to the headmaster.

In the staff-room before assembly Mr Beech approached me with some excitement.

'What's this I hear about your gander – George, is it?'

'Gregory.'

'Well . . . the caretaker tells me that Gregory is a hero. We had a burglar, and two policemen, a caretaker and a teacher all tried to catch him – but the actual arrest was made by your gander.' He was beaming with pride.

After the formal part of the assembly that morning,

Mr Beech announced that Gregory the school gander was a hero. There was a great burst of applause for him when the tale was told.

I smiled . . . and wished that Gregory could have heard it. I am sure he would have been mightily pleased.

Elderberry Wine

DURING THE last eight weeks of pregnancy, the ewe's foetus develops at a remarkable rate – prior to that it is relatively small. In these final weeks we gradually increase her ration of concentrates to two pounds a day, in two feeds. She gets noticeably larger and her udder begins to 'bag up'.

In 1980 we had purchased two champion Kerry Hill ewes at an auction in Welshpool. We called them Jemma and Jenny and they have produced lambs every year for us. They are both great-grandmothers now – and they have given us very little trouble, except Jenny – she worried us once. She was in lamb and progressing normally at first, but as her time drew near she began to grow at a disturbing rate, until she was so enormous she could barely squeeze through the doorway of the pen. Nevertheless I expected no problems on her lambing date – March 7 – for she had lambed five times before and it had always been like shelling peas for her.

'Jenny's gigantic, Sir – is she going to explode?' was a typical question from my concerned pupils. I had to admit that I had never seen a sheep as large as she was.

'How many lambs will she have, Sir?' was a common query.

'The size of the ewe is not always an indication that

98

she will have twins or triplets,' I would answer. 'Sometimes it means an enormous single lamb.'

The children began to have bets on the number of lambs Jenny would deliver, and I had visions of a record-breaking multiple birth – perhaps she would have a litter to match that of a sow or a rabbit!

Three days before she was due to lamb she went off her food and just lay in a big heap, like a monstrous woolly balloon.

'She'll be okay after she's lambed,' I told my uneasy pupils. 'The unborn lambs are too much for her. She's getting on a bit – she's a great-grandma, you know – and they're sapping all her energy.'

On March 5 – two days early – Jenny was on her feet and pawing the ground. She was doing all the characteristic actions . . . she was obviously going to lamb. It pleased me that she was two days early and keen to get on with the job.

It was early afternoon and I could easily keep an eye on her – our classroom is only a few feet from the lambing pens. I was feeling tired, having been up half the night with another ewe, who had produced twins. My torpor vanished at the prospect that Jenny might be even more successful.

After an hour of pawing the ground, her water bag appeared and burst. She started to strain, pushing really hard. But she seemed to be making no progress and I had to make a decision. Should I let her try for a while longer on her own, or should I investigate?

The class had gone for their afternoon break, and I was free for the final period so I would have no major interruptions. Matthew, my latest pupil to be keenly interested in sheep, was not concerned at missing break and he returned to check the flock.

He looked worriedly at Jenny. 'Is it warm water and soapflakes time, Sir?' he asked.

'Yes, please,' I replied. He reappeared in seconds with warm water, Lux flakes, and a towel.

He held the ewe. She was a patient old girl and I did not expect that she would struggle unduly. I rolled up my sleeve and lubricated my hand and arm with soapflakes. I could see no sign of a lamb under her tail, but just inside the ewe I felt a head.

Further investigation told me that the front legs were not in the normal position, but were bent back under the lamb. In between Jenny's strains I pushed it back into the womb. There was very little room to manoeuvre, and I could tell that there was another lamb in there.

Concentrating on the first lamb, I brought its legs forward into the correct position, pulled it gently, and out it came. I removed the membrane from over its nose;

it coughed and spluttered and began to breathe. It was a well-marked ewe lamb. I quickly placed it under Jenny's nose and she got up and began to lick it dry.

When she lay down again and began straining I sensed that all was not well. I soaked my arm again and discovered a mass of legs. Yes – there were two more lambs to come! I closed my eyes to help me shut out the world and concentrate. I found the rear end of one lamb, located its hind legs, and realised it was tangled up with another lamb. I could feel this one's head and its front legs, one of which was bent back. I could not separate them – they were interlocked.

I pushed them back into the womb to give me more room to work, then pulled out the lamb which was in the 'backwards' position. It was alive – a well-marked ram lamb. Jenny started to lick him dry. Now I brought out the remaining lamb, in a normal presentation this time. It was a live ewe lamb with very pale markings. Jenny did not quite know which one to lick first! I checked once more – no, there was not a fourth lamb inside her.

I suddenly became aware that I had an audience of ten pupils. It was now after school but they had stayed on, risking a telling off for being late for tea rather than miss Jenny's big event.

'Will she be all right now, Sir?' asked Matthew.

'Oh yes,' I replied breezily. 'The best medicine for her was to get rid of those lambs!'

Although his parents were unconnected with farming – his father was an electrician and his mother a teacher – Matthew was immensely keen on sheep and farming in general. He was learning fast and I was fairly sure that he would obtain seven or eight 'O' levels, go on to sixth-form college for some 'A' levels and read agriculture at college or university. His birthday and Christmas presents were generally books on farming.

As we proudly surveyed our new triplets we were joined by Phillippa, a fifth-year girl who had had to go to the dentist for a filling. Her parents were not farmers, either – her dad was a works manager and her mum a part-time helper at a primary school. She was determined to be a vet, but unlike others with less ability and a more fanciful view of the profession, she already had a fair knowledge of what it involved. She had been working part-time at our vet's surgery for the last twelve months and was a valuable helper on the school farm. She was in the top maths, English and science groups, and I was confident that she would be my first pupil to achieve her ambition.

'You should have been here sooner,' I told her.

'I can see I've missed all the action,' she sighed.

'Never mind,' I said. 'After you've qualified as a vet, you can come back and do these jobs for me . . .'

'I'd like that.'

'Without pay, of course,' I added.

'That would not be professional,' she declared. She had her head screwed on squarely, had our Phillippa.

By this time, after a few wobbly failures, our three lambs were on their feet. I pointed them in the direction of the 'milk bar' – Jenny was full of colostrum – and they each had a good drink. I injected Jenny with antibiotics to prevent infection.

I checked them all at 9 pm and again at bedtime. Jenny did not look her old self yet. She had drunk some water, but had hardly touched her concentrates and hay.

There was more action when I returned in the early

hours. Susie, one of our older ewes, gave birth to a single ewe lamb. That was fine – but then Tessa produced a large dead ram lamb. She was a good mother and full of colostrum, so I decided to foster one of Jenny's trio onto her.

The best time to foster a lamb, in my experience, is immediately the ewe gives birth. Over the years I have developed a successful stratagem for this. I take the dead lamb and rub it against the live one. Then I find the wet place on the floor where the waters from the intended foster-mother have burst. Scooping some of the fluid up, I bathe the live lamb with it – this procedure, I can assure you, stops me biting my fingernails and sucking my thumb!

I hide the 'gunge'-covered live lamb under my coat – yes, the coat I'm actually wearing (it would probably stand up by itself by the end of the lambing season). Then I place my hand inside the foster-mother and pretend to pull out a lamb from her.

I take the lamb from under my coat and move it from her rear end to her front. Usually she will start to lick it and make motherly noises. If this does not work I have skinned the dead lamb and placed this skin on the live lamb which is to be adopted – this usually works.

Tessa certainly took to Jenny's lamb – it was the largest and strongest of the three, a well-marked ram which I would hopefully show one day.

Jenny, though, was still not well. The vet had been to look at her; she had a temperature, and I began giving her a five-day course of antibiotics, along with some red medicine which I collected from the vet's. This was squirted down her throat with a syringe, and she hated the taste of it.

There was more bad luck to come. Another ewe, Trudie, gave birth to a dead single lamb. She was an excellent mother, and I had no hesitation in giving her both of Jenny's remaining lambs. She took to them in seconds.

Poor old Jenny was eating less and less as the days passed. When the vet called again to examine her he gave her calcium and magnesium, and I began giving her a multi-vitamin injection as well as the antibiotics.

After five days of antibiotics there was no improvement and she was eating nothing. The vet recommended another antibiotic, but four days later she was still no better, and beginning to look like a walking skeleton – except that she was not walking, and would only stand up for a minute or two when she was forced to do so.

The vet told me to keep on with the antibiotics, and not to give up – as 'where there's life there's hope' – but he did not think it looked too promising. She would probably die, he said.

That night, as I lay in bed, worrying about Jenny, I suddenly remembered something my old farm boss, Mick Hardy, had told me. I thought it went something like this: 'I never bother with them fancy expensive antibiotics, unless I really have to. If you get a ewe that's lost the will to live, get the old girl drunk on elderberry wine.'

According to Mick, the combination of alcohol and the goodness from the elderberries would perk her up and usually cure her – 'that elderberry has got healing properties, without a doubt. . . .'

Well, I was ninety per cent sure that this was what he had said. I toyed with the idea of phoning him to make sure, but I could imagine what he would call me for getting him out of bed after midnight – he certainly would not be inclined to discuss the merits of elderberry wine at that hour!

I got out of bed and checked the stock of home-made wines in the garage. There was plenty of peach, grape and wheat, but no elderberry . . . just my luck. I was about to go back to bed when I spied a solitary wine bottle on a top shelf in the corner. It was almost hidden by dust and cobwebs. I climbed up the stepladder, wiped off the debris, and managed to read the label. It was elderberry, all right, and from the date it was ten years old.

That might do the trick. Perhaps when I opened it a genie would appear and grant me three wishes. At this moment, my wishes would be: that the ewe would be cured; that I could go back to bed; and that someone would take over the sheep, making my life easier!

When I uncorked it in the kitchen, it smelt good . . . really good. It was no use giving the ewe wine that was 'off' and I decided to sample it. I poured out a small wineglassful and the smell – the bouquet – was delightful. The first sip was like nectar. I wanted more, and I soon finished the glass.

By the time I had driven to the school, less than a mile away, I felt extremely warm inside and my vision was playing tricks. Was it possible that I might fail a breathalyser test after merely one glass? Still, if I felt like this, the ewe would surely benefit. . . .

I poured about a quarter of the bottle into a clean saucepan and heated it gently on the boiler. The classroom was filled with an enchanting perfume.

When I showed Jenny the saucepan of hooch she would not – or could not – drink it, so I syringed it carefully down her throat. After the final syringeful her head flopped back onto the straw. I went back to bed, where I slept like a top.

At 5.30 am, when I next looked at Jenny, she was fast asleep. You do not often see that in a sheep; they tend to doze off in fits and starts. Sleep is a good healer, I thought, as I tiptoed out of her pen.

She slept profoundly until lunchtime, and when she awoke her eyes were distinctly bloodshot. The cackle of a nearby hen seemed to disturb her, and when one of the calves started to moo the expression on her face was positively pitiable. She appeared to be in need of an Alka Seltzer.

'Whatever's the matter with Jenny?' whispered Matthew.

'I think she has a hangover,' I said.

'You didn't take her boozing with you last night, did you, Sir? I know you like sheep . . . but isn't that taking it a bit too far?'

'I didn't take her out,' I assured him. 'But I did give her some elderberry wine, to see if it would cure her.'

'Well, it hasn't,' he declared. 'Just look at her!'

I had to admit that she looked awful.

At afternoon break the headmaster stopped me in the corridor.

'What's all this about giving your sheep elderberry wine, man?' he demanded.

'It's true,' I said. 'I think it's an old-fashioned cure.'

'Before you bring alcohol onto the school premises, you must have written permission from the governors,' he snapped in the stern tone he usually used for admonishing naughty pupils. I recalled that Mr Martin, the mathematics teacher, had suffered a similar ticking-off when he brought a couple of bottles of sherry into the staff-room to celebrate the birth of his first child.

'I'm sorry, but it was – and still is – a matter of life and death,' I countered. 'I only thought of it at midnight, and I don't think the governors would have been pleased if I had tried to get written permission at that hour!'

He walked away, muttering that he wished he was headmaster of a normal school, and leaving me wondering how he had found out. It amazes me how he does it!

Miraculously, at four-thirty that afternoon Jenny was back on her feet. She still looked like a skeleton, but now she was actually walking, and her eyes were no longer bloodshot. She drank half a bucket of water and, to my infinite joy, began to nibble some hay.

When I checked the flock that evening I decided to give her a further tipple. I warmed another draught from the bottle, firmly resisting a strong temptation to have a glass with her. Again I gently syringed it down her throat. This time she licked her lips in what seemed like gratitude.

The next time I saw her, at 3.30 am, I expected her to be asleep. But she got up as soon as I entered the pen, and walked over to me. Walked? It was more of a wobble . . . there was no way she could move in a straight line. She was plainly and happily drunk.

I switched off the light and left her to it. I just hoped she would not start singing and waking up the neighbours. . . .

The next morning she was lying in an immobile heap again. She was well into a second hangover, and again touched no food until well into the afternoon. Then she got up and ate some hay. Soon she was tucking into her bowl of concentrates – the first she had eaten for days.

At bedtime, I poured a smaller measure into the saucepan – there was less than half a bottle left and I wanted to keep it for future emergencies. This time there was no need for the syringe. Without even a sniff, she drank it up, straight out of the saucepan, and cheekily baaed for some more. I could not resist giving her another small glassful.

'That's your lot, girl,' I chuckled. 'You really are better – but I think you are rapidly turning into an alcoholic!'

She baaed as if to say: 'I don't care, just give me some more. . . .'

Jenny was soon back to normal. And her three lambs were doing well on their foster-mothers; two of them, I felt, might turn into show-winners.

I could not help bragging to Martin, the vet, about the astonishing elderberry wine cure.

When I saw him a week later I said: 'The antibiotics

and the medicine didn't do the ewe any good – but that wine did the trick, all right.'

'It wasn't the wine,' he declared. 'The antibiotics took time to work. Your elderberry wine might have made her feel better . . . but it didn't cure her.'

'I can't agree,' I said. 'You want to carry a case of wine in the back of the car when you go on your rounds. I'll make enough for the whole practice. I reckon I could sell it to you for three pounds a bottle.'

'Always the businessman!' he exclaimed. 'No . . . I'll stick to my antibiotics.'

I persisted: 'Well, you could always carry a few bottles around. Then, if you felt like a drink. . . .'

'I'll stick to whisky, thanks. Now look, John, stop kidding yourself. You could have given that ewe plain water, or flour, or sugar, or even bananas, and she'd still have got better. The antibiotics cured her – and it was just by chance that the wine seemed to work.'

I know different. There is nothing like a drop of booze to drive away a ewe's blues.

109

We Could Do with That Grass

IN THE September of 1985 I decided to give my new second-year pupils a taste of 'real' farming by getting them some calves to look after.

The prospect aroused tremendous excitement among the new intake. Most had had no contact with farming until they were introduced to Rural Studies and our school farm – of the 240 newcomers, in fact, only a handful had relatives who were actually involved in farm work.

Their only genuine experience of animals was with the pets they kept – cats, dogs, hamsters and goldfish were the main species. A few were fortunate enough to go pony-riding at weekends.

In previous years I had encouraged the new intake to care for our smaller livestock – rabbits, chickens and ducks. Now it was time for something bigger and more farm-orientated. I reckoned that a good target to aim at would be four calves, to be shared among my eight second-year classes.

At the start of the next eight lessons, I told each class of thirty about our great calf crusade. The big problem, I explained, would be getting the money to buy the beasts.

We raised the cash by buying plucked chickens, preparing them and selling them ready for the oven. In the end we sold five hundred chickens – and saved every bit of the profit. I phoned Dennis Goalby and found he had plenty of calves to sell. And after a little haggling, I managed to buy five, not four, Hereford × Friesian bull calves for ninety pounds apiece – well under their market value.

These calves, now the exclusive concern of my youngest pupils, were by no means the first to come into our school farm. We had started with our first two – Pinkie and Blackie – way back in 1974. We have now found that Hereford × Friesians suit our operation. They are not as expensive as the continental breeds, and we can easily find buyers for them after they are six months old.

We find it more difficult, though, to sell similarly bred heifers. We favour good-quality animals that will sell to farmers before they reach beef age. We have bought them from markets, but I have found that buying privately, from someone like Dennis, is more profitable. His calves are a fair price, bright and healthy, and guaranteed for a week. Calves from markets can bring trouble with them, such as pneumonia and scours.

I like to buy our calves on a Friday or Saturday evening. Then they have a couple of days to settle in before they are besieged by hundreds of admiring pupils.

My second-year pupils were pleasantly surprised to see five calves instead of the promised four. Another surprise was that one was red and white, instead of black and white.

When a Hereford bull is crossed with a Friesian cow, the dominant colour in the offspring is usually black – occasionally, however, a red-and-white calf is born, making an attractive contrast.

The pupils vote on the names to be given them, and

enjoy all the work they entail, both in and out of lessons. The calves are weighed each week, and graphs are constructed to plot their growth-rates.

They are only a week old on arrival, and are fed substitute calf milk; they are weaned at five or six weeks. Dennis had trained these five to drink very well out of the bucket, and they were all eager to drink – in the past we had sometimes had dozy slow-drinkers.

We mix the milk substitute with water, a cupful to four pints. The optimum temperature for the brand we use is 38 degrees Centigrade. The calves get the substitute milk twice a day, and to encourage them to eat solids we place a small quantity of dry food in the bottom of the bucket as soon as they finish their milk.

They graduate to solid food – a palatable coarse mixture of flaked maize, linseed, protein pellets, rolled oats and molasses – and hay ad lib. After eight weeks we gradually introduce rolled barley, and by twelve to fourteen weeks they thrive on a feed of 50 per cent of this and 50 per cent of the coarse mixture. All feed is weighed and the pupils learn to work out the conversion ratios. Eventually they draw up a balance sheet which includes the purchase of the calves, the cost of feedstuffs, bedding and any vet's bills – expenditures which are set against the sale of the calves.

We have no labour costs, no rent and no electricity to pay, and so the calves hopefully make a 'profit'.

I enjoy these practical mathematical lessons, and in my opinion they are much better than the theoretical ones.

Our famous five grew on strongly, and were ready for sale at seven months old. I would have liked to have kept them longer, turning them out for grass, but we needed the grass for our flock of ewes and lambs. Just over half of our one-acre school farm consists of grass paddocks, and it is obviously not enough for our breeding ewes which by now had reached twenty in number. We were forced to rent, borrow and take grass keep.

Ours is not the only grass in the school grounds – there are lawns, playing fields (including an excellent cricket pitch), and four large quadrangles. These non-farm areas are kept clipped by the county council.

I often asked the headmaster to let us graze livestock on them. He would not agree. 'What on earth would the county council think?' he would demand, adding: 'There's no chance of you taking over the playing fields – this school has a strong cricketing tradition, and it must continue!'

I did not really expect to get the cricket pitch . . . but I was certain that our sheep would be an asset to the football pitch, which could do with fertilising. And they could do no harm to the quadrangles, which were not used for anything. . . .

The spring grass was growing vigorously after a long, hard winter. I gazed enviously at the four quadrangles, around which the school's assembly halls, classrooms and corridors are built. For some reason the grass there had not been mown by the county council, and it was just about the right length for grazing.

I buttonholed Mr Beech. 'Headmaster, do you think we could put a few sheep on that lovely grass?'

'Certainly not, Mr Terry. I don't want the county council complaining that we're doing their job for them – why, the workers might go on strike!'

'I don't think they'd do that . . . after all, it would mean less work for them,' I argued.

Mr Beech was adamant. 'The sheep would make far too much mess,' he said. 'And when the council men did mow the grass, they would get sheep-mess on their shoes.' He obviously had not forgotten the time he arrived at an important county council meeting with goat-droppings plastered over the soles of his hand-made leather shoes!

'The sheep would disturb pupils and staff working in the main buildings,' he said with finality.

It was no good. I could not supply the sheep with rubber bungs to stop them making a mess – and gags to keep them quiet.

Next day I transported our ewes and lambs to the field, nine miles away, which we rented from Lord Clifton. But I maintained a covetous eye on the grass in the quadrangles. It responded to the showers and sunshine of a friendly spring by growing with a superb lushness, in spite of the fact that it had received no fertiliser. It was soon too long for our sheep, but it was ideal for silage – and in a little while it would make good hay.

For a moment I imagined myself bringing a tractor and some implements from Mick Hardy's farm to do the job. But I would not stand a chance of getting anything wider than a wheelbarrow through the small door into the quadrangles.

Perhaps I could get the council to mow it – then I could load it into our trailer and dry it somewhere in our Rural Studies area. I had plenty of labour, and I have made excellent hay by hand. If it looks like rain you can bring it in, half-made, and put it out again when the sun shines – not many farmers can do that!

I looked out every day for the council workers, but for some reason they did not materialise. Perhaps the grass had got too long for them to tackle.

One lunchtime the headmaster advanced on me: 'Have you seen all that long grass in the quadrangles, man?'

'I certainly have,' I replied politely. 'Do you remember, Headmaster, I asked you. . . .'

'Now look here,' he interrupted. 'I've an important meeting here at the school in a couple of weeks, with many headmasters coming in from other schools. I certainly don't want them to see it like this . . . will the sheep graze it off?'

I couldn't believe it!

'Actually, it's a bit late,' I told him. 'We've taken all the sheep to Lord Clifton's farm. In any case, the grass is really too long for sheep now – but it would be fine for our five calves. We were going to sell them, because they've grown too big for their shed, but market prices

115

aren't so good at the moment. If I could keep them a little longer, we should be able to get more for them.'

'All right, man – I don't want a lecture on agricultural economics . . . do you want the grass, or don't you?'

'Yes, please,' I enthused. 'I'll get the calves in the largest quadrangle within the hour.' I thought I would get the job in hand before he changed his mind.

'Splendid,' he said. 'I hope they're hungry – I want the grass to look like a lawn again, quickly.'

I did not have the nerve to say, 'What will the county council think?' – I just said I would make sure it was tidied up with our rotary mower after the calves had done their part. It was the least I could offer for all that lovely free grass. . . .

After lunch I marched out one of my fourth-year classes, which was my smallest group of only fourteen pupils, and hitched the trailer to my car.

'Are we going out for the afternoon?' asked Samantha.

'No – we're moving the calves into the largest quadrangle.'

'You're joking, Sir,' she said. 'I know for a fact that you've been trying to use that area since I started at this school.'

'Actually,' I said, 'I've been trying to get at the grass in those quadrangles since I started teaching at the school!'

'Did you make your first application in Latin, Sir?' quipped Robert.

I had decided to use the trailer because it would be too hazardous to drive our lusty calves all the way, and they would not lead on halters. I would have to drive them indoors for the final stage – through a large wooden door into the school's entrance hall.

As you enter the hall, there is a medical room on the left and the deputy headmistress's room on the right. Beyond them are corridors to left and right which would have to be blocked off. The door to the quadrangle was straight ahead. It would be easy!

We loaded the trailer with hurdles and backed up to the large wooden doors at the front of the school. There we constructed an unloading bay, and inside we used other hurdles to seal off the corridors.

The deputy headmistress, Miss Latham, emerged from her study to ask what was going on, and was very pleased that we had been given permission to use the grass. She had plenty of marking to do, she said, and she would make sure that her study door stayed shut.

It was not straightforward to get our first load – three of the calves – into the trailer. They had been in their pen for seven months and did not fancy leaving it. After much pushing, the biggest – Rambo – made a giant leap onto the trailer ramp, and the other two followed.

I drove the short distance to the front of the school, closely followed by my class, and backed into the unloading position with perfect precision.

'That's the way to reverse a trailer,' I boasted.

'Pride comes before a fall,' said Fiona.

I lowered the ramp and opened the gates on the back of the trailer. Robert tentatively banged the side with his hand. The calves did not budge. Robert tried again with a mighty thump which caused Rambo and his pals to hurtle out – straight into the school entrance hall.

When they were level with the medical room the door opened. Out stepped a nurse, carrying a tray of needles and syringes in one hand and a roll of cotton wool in the other. Rambo felled her instantly.

'Crikey, Sir, you've killed the nurse!' yelled Robert. 'She's been trampled to death, Sir!'

The calves were galloping around the entrance hall, playfully kicking their heels up. Robert helped me to pick up the nurse while Samantha and some of the others searched for the scattered needles and syringes.

The nurse's eyes were glazed and she was trembling violently as we bundled her back into the medical room.

She was clearly in a state of serious shock! I wondered whether I should offer to give her an injection to settle her nerves – but as my injecting experience had been limited to livestock I thought better of it.

We sat her down and she did not speak; she just seemed to be shaking all over.

'You'll be all right in a minute,' I said, in soothing tones tinged with panic.

'Didn't you know the second years were having their injections against tuberculosis today, Sir?' asked Fiona, sarcastically.

'N-now you mention it, I did see something on the staff noticeboard,' I stammered. 'But I'd forgotten all about it . . . it didn't really concern me.'

'It concerns you now, Sir. It's bad luck, isn't it? This nurse only comes here three or four times a year to give us our jabs.'

'She would come today of all days,' I said with a groan.

Suddenly there was a loud bang outside the room, and I instinctively knew what it was – a hurdle hitting the floor. I dashed into the entrance hall, just in time to see the backsides of the calves disappearing, way up the corridor we had blocked off. Horrified, I watched them turn into the English classroom at the end of the corridor.

I pelted after them with my pupils, and we all piled into the room, panting from the chase.

It was a rather odd English lesson. The second-year pupils looked up from their essays to see three calves prancing in front of the blackboard, heels kicking and tails high, followed by a bunch of red-faced, breathless, unruly-looking fourth-year pupils. The calves were out of breath, too, and one started to cough. The essayists giggled incredulously.

'What's going on, Terry?' thundered Harris, the English teacher.

119

'I'll explain everything in a minute,' I said, motioning to my students to round up the runaways.

'But . . . but . . . dammit, man,' he muttered weakly. 'I've got calves in the classroom!'

We drove them back along the corridor, treading carefully around a steaming patch of dung, and headed them into the quadrangle. They kicked up their heels and gambolled around. They loved the open space, obviously – it was better than their pen, or the entrance hall, or the English lesson. . . .

It was time for me to face the fireworks. What should I do next – comfort the nurse, placate Harris, or get the mess in the corridor cleaned up before the imminent lesson-change?

I instructed a couple of my pupils to run back to our department and get shovels, brushes, and a bucket of disinfectant. If the caretaker happened to come along and spot that mess, he would be fetching Mr Beech and threatening resignation. . . .

I went back to the nurse. She had calmed down and more or less stopped shaking. She must have been feeling better, because she started to tell me off.

'I don't like animals,' she said. 'I'm scared stiff of them at the best of times. How could you do this to me?'

There was a bruise on the side of her face, either from the fall or where a calf had kicked her.

'I've been kicked in the face,' she moaned. 'It hurts terribly.'

'I'm awfully sorry,' I said.

At this point Harris burst into the room with yet another tale of woe.

'Terry,' he said ominously, 'what are you going to do about the mess on my floor?'

'I hadn't noticed any,' I said.

'That's typical of you. You wouldn't notice, would you – you're as happy as a pig in it!'

I went into the corridor, where my pupils had begun to clear up, and instructed them to deal with Harris's classroom as well.

Harris, who had followed me, was not pacified. 'I still want an explanation from you,' he threatened. 'And it had better be good. . . .'

The nurse was checking syringes and counting bent needles, while a couple of my pupils tried to cheer her up.

'You ought to see a nurse for that bruise,' joked Fiona.

'A piece of rump steak would do it good,' suggested Robert.

'Rump steak is for black eyes,' she said, melting a little. 'Anyway, it was rump steak that caused my bruise in the first place.'

To me she said: 'You owe me for a new pair of black tights.'

Having to part with money was a bruising prospect for me at any time. Knowing this, one of my more playful girls quickly pressed the point. 'I'll go into town and get her some tights now, Sir,' she volunteered.

'All right,' I said. 'Here's a fiver. Make sure you bring me some change.'

It was now lesson-change – my chance to eat humble pie before Harris decided to make noises to a higher authority.

'That's the first and last time I shall have calves in my classroom,' he declared in his disinfected sanctum.

'Yes, I'm sure it will be,' I said earnestly.

'This must be the only classroom in the whole of the

British Isles to have had calves in it this afternoon.'

'I'm very sorry for all the trouble we've caused you,' I said. 'But you only know half the story.'

He shook his head and tut-tutted.

'The calves knocked the nurse over,' I went on, 'and her syringes and needles ended up all over the place. They knocked over the hurdle we put up to barricade the corridor – that's how they got down to your classroom.'

'You should have told the nurse you were coming – then she would have stayed in the medical room,' he glowered.

'I realise that, but I'd forgotten she was doing the injections this afternoon.'

I thought a bit of nostalgia might help: 'She reminds me,' I said, 'of the nurse who used to come when I was a pupil here. She was the "nit" nurse – she looked for lice or "nits" in our hair. We called her nitty Nora the bug explorer!'

Harris actually smiled at that! He had not finished, though.

'You should have made that barrier more secure,' he declared.

'Yes, but you should have had your classroom door shut,' I retorted.

'I suppose so,' he said. 'But I'm not taking the blame.'

'It's my fault,' I said, bravely.

We led the other two calves to the quadrangle on halters. It was the first time they had been led, and they were reluctant, but we soon had them safely in the quadrangle.

'Why didn't we move the first three like that?' asked Fiona.

'You live and learn,' I said dejectedly. I was all on edge – surely someone would have told the headmaster about the calf fiasco by now? The 'jungle drums' would certainly be throbbing. . . .

I suddenly remembered the deputy headmistress. I knocked on her door and waited nervously. 'Come in,' she said. 'Have you moved your calves? Can I come out of my office now?'

'Yes, we've moved them . . .'

'No problems, then.' Before I could say any more, she added: 'Very good, Mr Terry. I'll carry on with my marking. I'm glad you have had a good afternoon!'

I waited all afternoon for Mr Beech to pay me a fateful visit. He did not appear, and I did not see him the next day, either. I lay low, convinced that the headmaster's wrath would descend upon me at any moment. The weeks went by, however, and amazingly he never mentioned the episode. Those jungle drums must have had a breakdown. . . .

Harris was offered another post in a comprehensive school in a neighbouring county and so he left for those pastures new at the end of that term. Not that he would see many 'pastures' – he was a 'townie' through and through and did not enjoy or understand the countryside. I do not think it was me or the calves in his classroom which made him leave – but it may have helped him make the final decision!

It is funny but I miss him in a perverse sort of way – the school's English department is very normal these days – with very obliging and helpful teachers.

Fayre Game

THE WHOLE school was preparing for one of its occasional bouts of frenzied fund-raising. Cash was badly needed for all manner of things, from extra textbooks to new curtains.

The Parent–Teacher Association held a large formal meeting to discuss the situation. Mr Bell, the deputy headmaster, presided. After dramatically spelling out the desperate need for funds, he cheerfully called for suggestions.

As expected by all present, Mr Beech, the headmaster, was keen on another sponsored walk by the pupils. He had an obsession about sponsored walks – and I would not be a bit surprised one day to find him leading a sponsored walk to raise funds to sponsor an even bigger sponsored walk!

Happily, members of the PTA committee had lost interest in walking. And they were not over-impressed by the idea of a sponsored table tennis tournament, or a sponsored silence – or a sponsored cycle ride.

An atmosphere of frustration and gloom descended on the meeting. Then someone suggested a summer fayre and there was a rustle of excitement which changed to downright enthusiasm when someone else said: 'Why not a fayre with a country flavour?'

For once the entire PTA committee was in agreement.

I learned all this the following morning – I have been asked to join the PTA committee, but I just do not have the time to attend.

'Mr Terry,' said Mr Bell in a tone vibrant with enthusiasm and purpose, 'We . . . that is, the PTA committee . . . have decided to hold a summer fayre with a definite country theme to raise money for school funds.'

'Oh, really,' I replied, feigning nonchalance. 'And what has that got to do with me?'

'I am counting on your help, Mr Terry. After all, you are our link with the countryside at this establishment. You know so many people connected with the land. . . .'

'I suppose I do know a few,' I observed modestly. 'But tell me more about this fayre.'

'It will be held on the large school playing field on a Saturday in the middle of June. There'll be a lot of money-raising stalls and sideshows, and we would like outside organisations to bring in trade stands and exhibitions.'

'Where does the countryside come in?' I asked.

'Oh, organisations such as the Young Farmers' Clubs, National Farmers' Union, conservation and wildlife groups – people like that – would hopefully put on their own little shows.'

That sounded fair enough.

'Well, come on, Mr Terry – can we count on your support?'

'I am very busy,' I said. 'But they do say that if you want a job doing, ask a busy person. . . .'

'Oh . . . that's really encouraging,' smiled Mr Bell. 'Let's face it, Mr Terry – you *are* a popular member of the farming and country community. . . .'

'Now you're creeping!'

'Not really. We know that if we want something for nothing you're the person to get it for us. Goodness gracious, you have the reputation of being able to get blood out of a stone!'

'This reputation of mine . . . I think I'll have to try to lose it,' I murmured peevishly.

'Yes – but not just yet. Come on, Mr Terry, are you game to help us?'

'I will agree to be an adviser,' I said. 'But I would like the Rural Studies department to put on a large stand – perhaps with livestock in pens, a display of pupils' work, and a stall selling produce, such as eggs, goats' milk, fruit and vegetables.'

Mr Bell was delighted – 'Oh, yes, a marvellous idea, Mr Terry!'

'There's just one point, though. We have had to buy food for the hens that lay our eggs, and more food for the goats that produce our milk, and seeds to grow the vegetables. So I wouldn't be able to give all the money to the PTA fund. What I would do, though, is to pay them – say five pounds or so – for letting us have the stall on the field.'

'You don't miss a trick, do you?' bellowed Mr Bell in agonised admiration. 'However, in this case I think it will work splendidly. Carry on with it – I'll leave all the details to you.'

And off he went, in search of other fayre game.

I explained the plan for the Rural Studies department's contribution to the fayre to all my classes, impressing on them the need to ask their parents very nicely to donate goodies such as home-made jams, chutneys, pickles and potted plants for us to sell.

127

There was an excellent response, and some of the pupils also offered to set up the stand on the Friday evening and Saturday morning. As the big event drew nearer I was asked to do very little work by the committee, apart from telephoning two or three farming contacts whose help would be needed on the day.

Of course, I had plenty of work to do in organising our stand. Most of the members of the PTA committee, too, were concentrating on their special contributions – Jim Bradford was organising a welly-wanging contest, David Bottomley was in charge of a coconut shy, Arthur Cooke was busy collecting for an antiques stall and Joan Jackson had everyone baking for her cake stall. Other attractions to be featured included china, bottle and white elephant stalls, and competitions for guessing the weight of a cake and the name of a doll.

Trevor Shaw, the popular secretary of the PTA, achieved many successes with the letters he wrote to various organisations inviting them to participate. Among those who agreed were a local conservation group, the regional water authority, and groups specialising in fishing, shooting, archery, spinning, weaving, crocheting, basket-making and wood-turning.

There were to be meat and cookery demonstrations, a fashion show, and information stands for the Young Farmers and other country organisations. The main attraction was to be a large dray pulled by a team of shire horses – the contribution of a well-known Midland brewery.

The committee took no chances with the weather. If it rained on the day, the two large assembly halls would house some of the displays, and others would go into nearby classrooms, while our own stall and display would be set up in the Rural Studies classroom and we would just open up our buildings and let people visit the livestock in their pens.

If it was fine, however, everything would be assembled on the school's large playing field – the field I had coveted for years for grazing purposes. It was annoying to think that this one day would be the nearest I would get to using it for our school farm livestock!

Our stall and display would need a lot of hard work and organisation, and I decided to form a special committee of pupils to run it. This, I felt, would help to achieve a responsible approach to the task. In assembly, Mr Beech read a note from me inviting anyone interested in running the stall to meet me in the mobile classroom at morning break.

I did not know how many to expect, though I knew

that many pupils had already committed themselves to helping on other stalls, or to working on other tasks at the fayre. I was therefore most pleasantly surprised when nineteen pupils turned up at my meeting.

I explained that I wanted every one of them to be involved on the day. We would certainly need a rota for those serving customers at our stall. I explained, too, that we would elect a committee of six pupils who would take over the major responsibilities of the operation.

It would have been a lot easier for me to select my team from the regular helpers who came in to feed the livestock before and after school, and at weekends and holidays. Yes – they could have done it without any problems; but here was my chance to interest some 'new blood' in the activities of the farm. Perhaps they would be around to help us after hours when the fayre was all over!

I suggested that we should have a chairman, secretary, treasurer and three committee members for 'Operation Fayre'. This course of action was based on my many years of experience on Young Farmers' committees – I must qualify as one of the oldest young farmers in Britain. Forming a committee was as easy as falling off a log for me.

'What will each person have to do?' queried Joanna.

'The chairman will be in charge, and make sure that things are running smoothly,' I said. 'The secretary will be responsible for any paper work. The treasurer will of course look after the money. The three committee members will have other responsibilities and will probably have to do a fair amount of fetching and carrying. . . .'

'The dogs'-bodies!' Matthew interrupted.

130

'Not really. I'm sure it will be very responsible work,' I said earnestly. 'Now – any nominations for chairman?'

'Paul Jones,' sniggered one of the girls. I do not think she was worried about him holding the office – she just 'fancied' him. He was a good-looking, muscular lad.

'I don't want the job,' he declared.

'Not a good start,' I said. 'Let's have another nomination.'

'Kevin Beswick,' suggested Paul.

'Any more nominations?'

'Joanna Mander,' said Anne Davies.

I asked for other nominations. There was silence and a few blank expressions. Kevin and Joanna left the room.

'All those who want to vote for Kevin, put your hand up,' I requested. Six hands went up.

'Now – all those who want to vote for Joanna.' There were eleven.

I asked Matthew to call in the nominees. They sat in silence, all agog.

'Joanna has been elected,' I announced, 'and so Joanna's now the chairman.'

'How can a girl be chairman?' asked Andrew, a second-year pupil.

'Well, actually she should be called madam chairman, or chairperson,' I advised.

'She's a proper madam, all right,' chuckled Andrew.

'Any nominations for secretary?' I asked, pressing on. There was only one, and Gillian Oliver was elected.

'Now . . . what about nominations for treasurer?'

'I suppose *you'll* want that job,' said Matthew with a wry sideways glance.

'No, I don't want the job – this is a pupil committee,' I said. 'I'm just an adviser.'

131

There were four nominations for this post, and Richard Cooke was elected. The other three nominees – one of them was Kevin Beswick, the unsuccessful candidate for the chairmanship – were all elected as committee members.

Break was over and the pupils had to run to get to their next lessons on time. Still, they had received a brief initiation into the ways of democracy.

When the newly elected officers and committee met at morning break next day, Joanna was quietly efficient in the 'chair'. We decided to put on show a ram, a pair of ewes, two lambs, two calves, four pigs, a goat and some hens, ducks and rabbits. Six pens would be constructed from hurdles for the larger stock, and some hutches and runs would be moved to accommodate the smaller stock.

We would pin photographs of the livestock – including, of course, our prizewinning sheep – and other aspects of the farm onto some display boards. Tables would be used to display CSE projects. For me the most important section was the sales table, from which we would sell our own farm produce and the bottled delicacies donated by parents.

The committee began to feel its feet, and came up with several good ideas – among them a competition for guessing the weight of our largest pig.

'What will the winner get?' asked Joanna.

'If it's a lady it will be a night out with Mr Terry!' hooted Matthew.

'That's quite enough of that,' I said sternly. I signalled for order as Mr Bell walked up, all bright and breezy.

'How's everything coming along for the fayre, Mr Terry?'

132

'Fine, thanks . . . no problems so far,' I assured him.

'One thing's bothering me,' he said. 'Who can we get to do the official opening? We could do with a celebrity . . . but one with rural connections. . . .'

'How about the National Dairy Queen?' I asked.

'Trust you to think of her,' said Mr Bell. 'But who is she – and how could we contact her?'

'I'll find out – leave it to me.'

I did find out, but it was no good. She was already booked up for another engagement that day. I suggested to Mr Bell that I should ask our local Lord Clifton, who had helped the school out on many occasions – he was letting us use one of his grass paddocks to graze our livestock.

On the Friday evening before the Saturday fayre we

transported our most neatly painted hurdles to the main playing field and set up our pens. We tied hessian sacks on the hurdles assembled for the pigs. I explained to the pupils that they would settle down better if they could not see out; if they were able to look through the bars they would probably try to tip the hurdles over and escape.

We thought it would be sensible to put the display boards and tables out early next morning, when we would have plenty of time to load them up with the motley collection of jams, pickles and plants now stored all over our classroom.

Fayre day dawned brightly. The forecasters promised it would be fine and warm all day – the PTA committee had chosen well. The event had been advertised in the local newspapers and all the pupils had taken home a newsletter urging their parents to attend.

I met the pupils early and, after feeding the livestock, we set up our sales table alongside the hurdle pens. We moved the sheep first, leading them on halters. That was easy – they were used to it.

Originally we had planned to take Anna, our oldest goat – she was quiet, good with children, and fond of fuss and attention. When we studied her with the fayre in mind, however, we realised she was looking her age. We thought the world of her, but she was definitely no longer a beauty queen!

One of her daughters, Ruby, seemed a better choice. She was a bit on the wild side, unlike the placid Anna, but she was in her prime, smart and shining. She travelled to the field in the trailer, with two small calves for company. Then we loaded up an assortment of hutches and pens to house rabbits, bantams and ducks.

Pigs were the final load. Joanna and I weighed the largest and marked it with a big cross. We vowed to keep the weight secret between ourselves until the competition was closed.

Once they were released into their new pen on the school field the pigs began to plough up the ground with their hard snouts. They had not been on grass before, and they loved it!

Other stallholders were setting up ready for the start at 2.30. The borrowed loudspeaker system crackled and hissed and Mr Bell's voice boomed out.... 'Testing, testing, one, two, three' in the strange ritual that seems to be compulsory at this kind of event.

The temperature climbed steadily. It was going to be a glorious day. The officials were waiting at the gates to collect the entrance fees and suddenly the local brass band launched into a rousing march. Incredibly, we seemed to have created the authentic atmosphere of a small agricultural show.

Now the public were drifting in, and we were waiting in frustration – we could not start selling until the show had been officially opened. But Lord Clifton had arrived, and was meeting the chairman of the governors. The shining-coated shires had arrived too, evoking shouts of delight and admiration as they surged onto the field in front of their impeccable dray, coats, harness and brasses all bright and shining.

Right on time, the deputy head's voice came over the loudspeakers – loud, but not very clear – to announce that his Lordship would now open the fayre. I joined some of my pupils near his makeshift rostrum as his familiar Eton accent came over with much more distinctness.

He began by apologising for not being the National Dairy Queen and for being 'second choice' as opener – however had he found that out? – and went on to talk in glowing terms about the school.

He mentioned our excellent academic standards, moved on to our sporting achievements and then congratulated us on our successes with the sheep at agricultural shows. Finally he declared the fayre open – and just as he said the word 'open' Richard tapped me on the shoulder.

'What are you doing leaving the money unattended?' I demanded of our stall treasurer.

He was so upset that for a second I began to suspect that our 'float' money had been stolen.

'Sir . . . Sir,' he spluttered. 'It's the goat. Ruby has got out of her pen, and we just can't catch her!'

Someone in the crowd overheard this, and began to shout, at the top of his voice, 'The goat's escaped!' – as if a dangerous circus lion was running amok.

On the rostrum Lord Clifton relayed the news over the speakers. 'Don't panic,' he announced with mock gravity. 'The goat's escaped. Keep calm, everybody.' Then he burst into laughter, and a mighty chuckle arose from the crowd.

I was beginning to look a right fool, but there was no time for worrying about that. Racing after Richard, I saw that Ruby was being chased by half a dozen of my pupils.

They were going about the task in quite the wrong way; their frenzied chasing merely made her run even faster.

'Stand still!' I shouted. 'Don't chase her, and she should stop running.'

136

They froze. Ruby ran a little further, then stopped, breathing heavily. Someone in the crowd made a grab and missed – she kicked up her heels and ran a short distance. I knew it would be almost impossible to catch her with all the people around. There was nowhere to corner her, and she seemed to know it – there was mischief in her eyes, as if all this chasing about was a tremendous game.

I had to try another tack, and I sent one of the girls to fetch some feed in a bucket. Perhaps I could coax her with that . . . but as I waited something in the crowd alarmed her again and she was away, heels kicking, udder swinging.

Townspeople know little about goats. Frightened folk were dodging out of Ruby's path, a little boy dropped his ice-cream and began to cry. My heart went into my mouth as Ruby gave a mighty buck and narrowly missed knocking an old lady over.

Now some younger boys began chasing her and I could suddenly see it all ending in tragedy. Some secret reserve of schoolmasterly discipline stopped me cursing them in most unprofessional terms; I heard myself calling, politely: 'Stop chasing her, please.'

I had to try to keep calm. But the children had not heard me, anyway. They were gleefully locked into the hunt, so it seemed – and Ruby, in full flight, was heading straight for the china stall.

What made those young hunters suddenly call off their chase, and why Ruby stopped so dramatically, within a yard of the stall, I will never know. I was certainly too shaken and frantic to ponder such mysteries at that traumatic moment.

A bucket of feed and a collar and lead were thrust into

my hands. I rattled the bucket. Ruby began walking towards me. She sniffed the bucket – I was almost touching her.

I could sense that I was being watched closely by an enormous crowd. I wanted to get it right. I *must* catch her now!

I was preparing to put my arm around her neck when a man in the crowd bawled: 'You can lead a horse to water – but a pencil must be lead!'

This utterly absurd statement drew a great roar from the crowd, and Ruby was galloping off again.

'Graphite, actually,' shouted Andrew in disgust to the joker in the crowd. 'Technical drawing is my subject, and I should know!'

It was pure surrealism. And it was lost on me.

'We haven't time for stupid arguments,' I snapped.

Ruby was heading back to the china stall. Its two lady attendants, ashen-faced, were bracing themselves for a grand disaster, with hundreds of plates, cups and saucers flying through the air and burying them alive.

Then, when the rampaging goat was little more than a neck's length from the nearest crockery, they waved their arms and shouted in unison: 'No! No! No!'

In the circumstances it was a wise and courageous act on their part, and one which seemed to be duly acknowledged as such by Ruby herself, for she stopped and turned in a flash.

As she turned I dived towards her and somehow got both arms round her neck. She was beaten. I clutched her in a grip of iron, and there was an exultant cheer from the crowd as I slipped the collar over her neck – a cheer which I could only recognise with a slight, grim nod.

She allowed me to lead her quite easily back into her pen, which was now doubly secured.

'This really is shutting the stable door after the horse has bolted,' I said. We were both panting, Ruby and I. But I calmed down after a while and began to think of our stall.

'How is business?' I asked Gillian.

'Very good, Mr Terry. But I'm sure we would sell even more if you'd stop trying to act like Clint Eastwood at a rodeo and give us the benefit of your superb sales techniques. After all, you've spent the year drumming the three Rs into us – reading, writing and retailing.'

'Well, there can't be many teachers who do that!' tittered one of our customers.

Our competition for guessing the weight of the pig was doing very well. For ten pence, entrants put their guesses into our book. Some were guessing in kilograms, others in pounds and ounces, and others in stones and pounds. We had weighed him in kilograms. It would be a good exercise for my pupils to do some converting into metric when we found who was the winner.

Mr Beech brought Lord Clifton to inspect our stall and he enjoyed chatting with the pupils. In fact, Lord Clifton knew some of my older pupils by their first names, and they liked that, from an actual Lord. He was kind enough to spend a small fortune on our stall. I could not help thinking he was taking coals to Newcastle, so to speak!

While Lord Clifton was making his purchases the headmaster drew me to one side.

'Mr Terry, I'm not being petty or argumentative,' he said, 'but why is it that it doesn't matter what we do . . . it could be an organised function like today, or some other less formal affair . . . but – *always* – at least one of your animals escapes and makes us look silly!'

'I wouldn't say that, Headmaster,' I said soothingly. 'They don't *always* escape. . . .'

'I must disagree, Mr Terry. Let's look at your track record. There was the pig that escaped and got into the home economics room – just as I was sitting down to tea with the school inspectors. Then there was that ram of yours – what do you call him, James Bond? – that got out of the field and served two of Mr Finney's ewes. Then the two sheep that were returned to the school in a police car, of all things. . . .'

'I can explain . . .' I began. But I could not get a word in edgeways.

'Let me finish, man. That gander of yours – George, is it?' – 'Gregory, actually' – 'Well, he escaped and chased

Miss Warner the needlework teacher and frightened her half to death. I keep asking myself: what on earth will escape next?'

'Oh, I'm sure this'll be it. By the law of averages, we shouldn't have any more escapes. You can bank on that.'

'I hope so, man,' said the headmaster darkly. 'For your sake.'

He was right – our livestock were certainly notorious for getting out and causing spots of bother. Mr Beech, happily, did not know of all the instances. I could have given him a few more – quite a few more.

I did not, though. I was not that silly.

At 4.30 the deputy headmaster announced the winner of our 'guess the pig's weight' contest. Five minutes later the lucky lady came to collect her prize.

She was an enormous woman who knew nothing about pigs, but she had guessed the weight exactly. Funny, that. Lord Clifton's guess had been miles out – and he is, of course, a full-time farmer.

'Where's my pig, then?' demanded the large lady.

'Well, your prize isn't a pig, actually,' I said with an apologetic smile. I pointed to the sign which stated that the winner would receive a box of chocolates.

'But I want a pig, for the freezer,' she said, somewhat belligerently.

'I'm sorry, but it does say a box of chocolates.'

'I'm not happy,' she said.

I pushed the box of chocs into her hand, feeling rather grateful that I did not have to conform with Matthew's idea that the winner, if a lady, should qualify for a night out with Mr Terry.

The crowds were going home. We decided to begin taking the livestock back to the school farm. Then there would be the long job of moving the tables and tidying up.

We had sold right out of goat's milk and eggs. The jams, chutneys and pickled onions had all gone, too. The few plants left could go back into our greenhouse.

In all the excitement I had not even had time to take a quick look around the fayre. I had missed the cake stall, the sideshows, the trade stands, the exhibitions. What a day!

We took our stall down and everything was all right but for one thing – the turf where the pigs had been penned was no longer turf. It was like a miniature ploughed field. At least the pigs had not escaped. I made a note to level the ground – the turf had vanished, but I could sow some grass seeds there and hope for the best.

Richard, our treasurer, counted the takings. After deducting the five pounds for the rent of the stall, I was delighted to find that we had made an excellent profit – a useful sum which would be used to buy hay for winter feed.

On the following Monday it was announced that the school had made a tremendous profit on the day. Some of it would be used to buy proper amplification equipment, to be put to good use at future outdoor events.

During the morning break I was approached by Mr Martin, the head of mathematics, and learned that he was very keen on hockey.

He pointed through the window at the school field and gazed upon me with utter loathing.

'Part of my hockey pitch is ruined,' he said. 'Your pigs have made it look like a bloody allotment!'

'Oh,' I said – a conciliatory murmur. 'I didn't realise that that was the hockey pitch. Tell you what, I'll sow some more seed on it and get it back to normal.'

'If you would . . . and be quick about it.'

He strode off, leaving me waving my arms helplessly and observing, to no one in particular: 'Animals . . . how could anybody be daft enough to keep them!'

The Challenge

THERE'S ONLY one topic of conversation at a Kerry Hill Society meeting – Kerry Hill sheep. And two questions loom large: 'What sheep have you got to show – and how good are they?'

It is my experience that most exhibitors do not tell you exactly what they have at home. Many of them will exaggerate, in this fashion: 'He's the best ram I've ever bred . . . in fact, the chairman of the society came out to the farm yesterday, and he says he is the best example of the Kerry Hill breed he has ever seen. He's nearly as big as my young Charolais bull – he's got enormous length and tremendous depth, with marvellous black and white markings and good upright ears.'

This type of hype gets you worried. You may even feel like carting your show stock off to the abattoir. A likelier danger is that you do not try as hard – you may cut back on food and skimp on carding and trimming.

Then this amazing ram arrives at the show, and you find it is nowhere near as good as yours, and stands 'bottom of the line'.

Another common brand of exaggeration – put out by a cannier rival – is that he has 'nothing much' to offer. It encourages you to step up your feeding and to spend hours and hours to get your sheep's coats looking immaculate.

144

At the show, however, you find that the so-called 'middling' sheep are as big as donkeys, with superb markings and ears like church spires.

'I can't understand it,' their owner will say. 'They looked so small at home . . . perhaps it's the large new building they're in.'

I heard plenty of tales of this kind at a Kerry Hill Society meeting when we got talking, after a few pints, about the forthcoming 1986 show season.

'Are you going to get another championship at the Royal Show this year, John?' asked Mr Stone.

'That's my ambition,' I said. I was well oiled by this time, and full of talk as well as drink.

'Our ewes were the female champions in 1982,' I bragged. 'In 1983 we had fourth prize, second in 1984 and third in 1985. So 1986 must be our turn for first prize and female champions. Besides, when we won in 1982 one of my rivals was heard to say: "It was just luck that John Terry and those schoolkids won." I vowed then that I would prove him wrong!'

I took another long swig of bitter. 'I've tried so hard for the last three years, and failed,' I added. 'I'm really going to try harder this year.'

'It does take some winning, though, John,' said Mr Stone. 'I know Mr Porter has a good pair of ewes this year. And without being big-headed, I think I've got the best pair of ewes I have ever bred – and I've been keeping sheep all my life.'

Here we go again, I thought to myself. . . .

Mr Williamson joined the debate – 'How many prizes did you win last year, John?'

'Oh, about sixty,' I said.

'That was quite good going, but nothing to write

home about. Do you know of anyone who's won a hundred in a year?'

'No,' I declared. 'That would really take some doing . . .'

Mr Williamson leaned forward with a sly smile.

'I don't know of anyone who's won a hundred – except me, of course!'

I sat upright sharply, almost knocking over my pint.

'That was in 1984,' he said.

'I bet we could get a hundred!' I blurted out the words without thinking.

Mr Williamson shook his head, smugly.

'I don't think you could . . . not with your small flock. How many have you got now – thirty, is it?'

'Actually, we are now up to forty-six,' I said – 'that's including all the ewes, lambs and rams. . . .'

'And goats and rabbits!' Williamson hooted.

'I'll bet you now,' he said gravely, 'that you can't win a hundred prizes in the year.'

'Oh yes we can,' I said. I drained my glass.

'I'll bet you a fiver.'

'Done!' We shook on it, and as there seemed to be an 'atmosphere' brewing, we changed the subject to the finer points of wool quality.

Next morning I realised what a fool I had been to take up the challenge. The money did not worry me – it was my 'pride' that was now well and truly on the line. I really needed to concentrate on my main ambition, another championship at the Royal; now I had lumbered myself with an additional burden!

We normally attend fifteen agricultural shows a year, covering some three thousand miles in the process. Would they be enough to produce a hundred prizes?

It was extremely doubtful – though making more entries in each show might produce more awards. In show schedules you usually find a Kerry Hill class for a ram two years old or over, one for shearling rams (they are one year old), one for ram lambs, one for a pair of ewes and one for a pair of ewe lambs. At a few shows you also get a class for a pair of ewes which have reared lambs that season. At some shows, there is just a class for rams of any age, rather than two-year-olds and shearlings. Some shows offer no class for a pair of ewe lambs.

At local shows, where there are no specific classes for Kerry Hills, the classes are for any breed of rams, ewes, etc. 'Any other breed' classes are scheduled when they have classes for breeds such as Suffolks, Texels and Jacobs.

This year we had two two-year-old rams to show, three shearling rams, one ram lamb, two pairs of ewes and a pair of ewe lambs. Our ewe lambs were nothing

much to look at, but I felt the others were good enough
to win a prize or two.

It would be the biggest show team we had ever
mounted – but I still could not see us getting a hundred
prizes with it. . . .

We spent a great deal of time halter training the show
team and six weeks before the first show of the season –
the Shropshire and West Midland at Shrewsbury – we
shampooed them all. As usual, we carded and trimmed
them each week, and on the day before judging we
washed their faces and legs.

This first show is very exciting – the other Kerry Hill
exhibitors cannot wait to see what you have in your
trailer, and you cannot wait to form a mind's-eye
judgment of their stock. I like to get there on the evening
before judging morning so that the risk of accident is
reduced and the sheep can get settled in. Then I go
home.

Next morning, after a poor night's sleep – during
which I usually dream of winning, or of my sheep
escaping and causing havoc! – I am up at 4.30 to meet
three or four pupils at the school gates. We quickly feed
the livestock, milk the goats – and get on the road.

I took three pupils – Matthew, James and Vanessa –
to Shrewsbury. By the time I had given the sheep a light
card and trim it was time for judging. We had five
entries and won five prizes. Only ninety-five prizes to go!

I was a little disappointed, though. We were first to
enter the ring with our ewes, Linda and Lisa, and at that
moment there was a tremendous downpour, and they
were thoroughly soaked.

Mr Porter held his ewes under cover – the judge asked
for him twice and he made some excuse about a missing

halter. His ewes were fresh and dry, for the rain suddenly ceased when he came out, whereas ours were sodden, and their ears looked a bit droopy. It was Mr Porter's ewes which won the female championship; ours were reserve champions.

Before the judging, at least six people told me how superb our ewes were – the best in the show, they declared! Afterwards, many more people, including some well-known judges, asked me what had gone wrong. Whatever it was, they had not won – and when I congratulated Mr Porter, I felt more and more determined to do better at the next show.

If we averaged five prizes at the next fourteen shows, however, it would mean only another seventy prizes. . . .

'We aren't going to do it, Sir!' exclaimed a disconsolate Matthew.

My £5 challenger, Mr Williamson, came up with a smirk to ask: 'How many prizes did you get, John?'

I wanted to retort, 'More than you.' But I suspected he had won six, so I said nothing.

I fetched the sheep back after school on the Thursday evening, carded and trimmed them until late on Friday, and set off for the Montgomeryshire Show at Welshpool at 5.30 am on the Saturday, with Elaine, Gail and James. We were the first to arrive with Kerry Hill sheep. As usual we tied clean bags on their pens and I gave them a light card and trim.

The judge, Mr Fred Johnson from Church Stretton, was soon ready to start, and we managed to secure two thirds and a fourth with the rams. In the ewe class we had entered two pairs. James and I showed our best pair, Linda and Lisa, and the girls – Elaine and Gail – took Lorna and Louise into the ring.

Robert Porter was again last to come into the ring. He automatically stood with his ewes at the top of the line, and I noted that at least it was not raining today.

I could see that the judge was impressed with the pair James and I were holding. He told us to stand them next to Mr Porter's, and inspected both pairs closely. He stood back for a different view, then told us to put our pair at the top of the line. After checking the other entries he signalled to the steward to hand out the rosettes.

It was our first 'red one' of the season. Our other pair were fourth.

Now the winning pair of ewe lambs was put alongside our winning ewes to determine the champion pair of females. It was ours!

During the afternoon we proudly took our cup-winners into the grand ring for the parade of livestock, which included shire horses, ponies, sweet little Shetlands, and plenty of cattle and sheep. With the lovely hills as a background, it made a truly magnificent scene.

At this show we had managed to capture six prizes – an improvement, but not good enough if we were to reach our target.

We studied the show calendar and found two additional shows which did not clash with those on our original programme – Pantydwr in mid-Wales and Moreton-in-the-Marsh.

They would involve an extra 380 miles of driving, but they would give us a grand total of seventeen shows, and we needed every show we could manage if we were to get anywhere near to our 'century'.

Our next show, two-and-a-half weeks later, was also

one of the largest – the Three Counties at Malvern. We had won the female championship there in 1982.

Could we do it again with our Montgomeryshire winners? I spent the weekend carding and trimming our team, and took them to Malvern after school on the Monday evening.

'Here comes the man that will have the second-prize ewes,' was the greeting I received from Robert Porter.

He went on: 'I was so disappointed that your ewes beat mine, John. Because of that I've spent hours getting their coats looking superb. I really am pleased with them, and I think mine will win tomorrow. Sorry to disappoint you, John!'

For a split second I felt like loading ours back into the trailer and going home. Still . . . he had not unloaded his sheep yet.

'I see you've bought a new trailer,' I observed in an attempt to be friendly.

'Yes . . . it cost us a lot of money. It's brand new, you know.'

He unloaded, to discover that one of his ewes had rubbed against the side of the trailer. A large greasy mark was running along the length of one side.

When Robert saw her the air turned blue with his wrath.

'The firm supplied it like that,' he spluttered. 'I never thought to wipe it out – there shouldn't be any need to. That bloody firm . . . I'll kill the sales rep!'

He snatched a cloth from the Land-Rover and tried to wipe away the offending mark.

Now it was not up to me to say anything – after all, he had been showing sheep a lot longer than me. But it would have been more sensible to let the mark dry and to card and trim it next morning. The wiping with the cloth only made it look a lot worse. Still, he was a very experienced showman, and I expected the ewe's coat would be perfect by judging time.

I picked up Vanessa, Matthew and James next morning and on our arrival at Malvern we gave our sheep a small feed and a light card and trim. Heavy rains had given way to bright sunshine by judging time, and I was pleased to see that the judge was Mr Beaver from Montgomery, who had given me much useful advice at previous shows.

Our two-year-old ram, the first to be judged, came second – an improvement on his thirds at two previous shows. We had two entries in the shearling ram class, and they were placed third and fourth.

Now our ewes were put to the test. And much to Robert Porter's disappointment, our Linda and Lisa were sent to the head of the line. Mr Porter stood next to them with his ewes, and our Lorna and Louise occupied the third slot.

The judge told Vanessa and Matthew to let them off their halters, and gave a similar command to Mr Porter. When the ewes ran loose in the ring, there was no doubt which pair looked the best, and the judge told Vanessa and Matthew to move ours up into second place!

So Mr Porter was down to third, with Mr Williamson fourth. Mr Porter told me wryly: 'I said you'd get second with your ewes. . . .'

Both of our pairs were brought back into the ring for the female championship. Linda and Lisa duly won it, and I expected that the reserve females would be the winning pair of ewe lambs – a class we had not entered. But that award went to our ewes, Lorna and Louise.

Our champion females were then put with all the other first-prize winners, including an excellent ram belonging to Mr Stone. This ram had been the champion Kerry Hill exhibit twelve times in a row, and was out to chalk up a thirteenth victory – rams usually get the supreme championship in Kerry Hill classes.

The atmosphere was positively electric as the judge stood our ewes next to Mr Stone's ram. I could hardly believe my eyes when he signalled to the steward. We had won the supreme award!

Our trophies that day included a cup, a piece of engraved glassware and a Three Counties necktie. Our prize tally was: supreme champion, female champions, reserve female champions; one first, two seconds, a third and a fourth. Eight prizes . . . making a total of nineteen for those first three shows.

153

A few days before the Royal Show, Andrew Jordan of Central Television called to say they would like to repeat a project they had done with us in 1982. They would film us at school preparing sheep for the show, and add it to further shots taken at the big event.

In front of the cameras, I carded and trimmed William, our three-year-old ram. Then Vanessa gave him a trim. We washed the sheep's faces, showed how we trained them to walk on halter, and finally loaded sheep into the trailer and drove off as if we were going to the 'Royal'.

I spent the whole of the Saturday and Sunday prior to the show carding and trimming. I was determined to reclaim the female championship which had eluded us since our triumph of 1982. I would prove, I vowed, that it had not been just luck!

My determination was shared by Adam, a former pupil who had been my best-ever sheep handler. He had helped us to win in 1982, and now he had come out of 'retirement' in the hope of doing it again.

I took the sheep to the Royal showground at Stoneleigh on the Sunday evening. I was teaching all day on the Monday, but Mr Stone fed them for me in the morning and I returned in the evening to get ready for judging day – Tuesday, July 1.

The judge was Mr Tom Willis from Rhayader. We were just about to start showing when Godfrey Brown, of BBC 'Midlands Today', arrived with a film crew, which followed us into the ring to focus on our rams winning two thirds and a fourth.

Then it was the ewes' turn. They stood perfectly. Adam and I were too nervous to speak. The cameras rolled as the judge passed along the pairs of ewes. After an eternity of suspense he told us to take our best pair to the top of the line. There was more delight; our other pair of ewes, handled by Vanessa and Matthew, were put into second place. Robert Porter's ewes were third.

Later we were presented with the prize I had been dreaming about over the last four years – the female Kerry Hill championship, on top of which the reserve championship went to our other two ewes. We had proved our point – our 1982 victory had been more than just 'luck'.

Going for the Hundred

A WEEK after the Royal Show the school closed for the long summer holiday. It would be a hard-working holiday for me though. As well as school paper work and the day-to-day running of the school farm we had many shows to attend. All sheep are carded and trimmed before each show and I had hours of work in front of me.

We had a problem with our next show outing, the Ashby-de-la-Zouch Show – a small one-day event at Ticknall, near Derby. The problem was that I must be at the Royal Welsh Show at Builth Wells by 9 the same evening as it was the rule that all livestock had to be on the showground by that time.

I expected to leave the Ashby show at 4 in the afternoon, which allowed me only five hours to get back to school, change some of the sheep over, feed the livestock, have some tea, and get to the Royal Welsh showground.

After clocking up a first and three seconds at Ashby, I left at ten to four and got to Builth Wells at ten past nine – just ten minutes late – after a desperate 140-mile drive. I took a three-year-old ram, two shearling ewes and our 'best' pair of ewes with me.

I always like to park on the quietest side of the exhibitors' parking area at Builth, because I sleep in the car on site as it's easier to take care of your sheep, and better than an hotel or bed-and-breakfast place, where you have to fit in your needs to suit the owner or manager.

The stockmen's accommodation on the showground is not for me – I would have trouble sleeping there, because it is in the middle of the showground, and these are noisy places at night. And, of course, sleeping in the car costs nothing and, as is well known, I am all for keeping costs down!

I managed to get plenty of rest on the Monday. Early on Tuesday, one of my keenest pupils – Vanessa – arrived with her parents. She was going to help with the showing and stay at a nearby aunt's house.

We worked on the sheep and it was soon time for the judging by Mr Alf Boulton, of Llanidloes. Our rams achieved two thirds and a fourth. Could we now win the championship which had so far eluded us at the Royal Welsh, leaving us with reserve female championships on two previous occasions?

Our ewes were the first to go into the ring, and they made a good impression, standing straight and square with ears pricked forward. We had them well trained – we could move well away from them and they would still stand like rocks. They easily outclassed the other ewes and were awarded the Cerney Cup as champion females.

At the Royal Welsh the supreme championship is awarded to a single sheep. As our best ewe, Linda, stood in line with the other first-prize winners I knew we were in with a chance. She was even more alert than usual, because she was looking for her show-ring pal, Lisa. She really looked quite splendid, and she won – the John Brick Cup for the best Kerry Hill exhibit in the show.

We did not stop there – we then went on to win the

157

group class, for a ram and two ewes shown by one exhibitor. Ours was a well-matched group of three large sheep with very distinct markings.

Our outstanding successes at this show were no small achievement. This part of Wales is the home of the Kerry Hill sheep. For a schoolteacher and a schoolgirl from Warwickshire to win three major awards amounted to a spectacular feat of showdom. We would be proud to include those distinctive Welsh rosettes and certificates, printed in two languages, on our school farm display boards. . . .

So far it certainly seemed to be 'our year' at the shows. However, we were still a long way away from our target of a hundred prizes.

We had some successes with our ewes at a local show at Shustoke, then pressed on in the hope of more at Burwarton, near Bridgnorth in Shropshire, a large show with Kerry Hill classes.

Our rams did quite well there, but our ewes again swept the board – taking first prize in their class, champion females, and supreme champion Kerry Hill exhibit. With our ram William, they also won the cup for the champion group of three.

We had won both cups on offer in the Kerry Hill section, and our tally was now fifty-two prizes from eight shows, an average of six and a half per show. We needed another forty-eight prizes in the remaining nine shows – or 5.3 prizes per show.

That sounded a cinch, judging by our success-rate so far. But many of the forthcoming shows were small affairs, often with no fourth prizes to offer. Usually they had merely a special prize for 'champion', and no female or reserve championships. It was going to be touch-and-go in our mission to reach those magical three figures. . . .

It was a local Saturday show next, at Canwell, near Birmingham, and we were well into the summer holidays. Our ram, William, did well in the 'other breeds' section, winning cups as champion ram and champion exhibit.

Next day we were at another local show, Fillongley. The following Thursday we headed for Guilsfield, near Welshpool, and the Saturday after that we were at Pantydwr – 140 miles away, and really too far to go there and back in a day. But we needed the prizes, and we were awarded another seven there.

159

At the Kenilworth show – part of the Town and Country Festival at the National Agricultural Centre – we won ten prizes. We were now up to eighty-eight, and we needed only twelve more at four shows.

'We'll do it easily!' bubbled Vanessa.

'We should. . . .' I replied, with reckless enthusiasm.

The summer holidays were over. The day after we started back to school, we were on the trail again, to Monmouth, and misfortune. Our rams and ram lamb could not get in the prizes. Talk about pride before a fall!

Our ewes saved the day, giving us a first, the champion females and the reserve champion Kerry Hill exhibit. They were substantial awards, and we had three to add to our total. We tried to look on the bright side.

Mr Williamson told us he would not be turning up at Knighton, the next show. It was one of his little jokes. He did turn up and his sheep snaffled about eight prizes. We again won only three – a first, a second, and the champion female Kerry Hill cup.

'How many today, John?' enquired Mr Williamson.

'Three.'

'That's not good enough if you want to get your hundred. Would you like me to give you some of mine?'

Before I could answer he said: 'You want to buy a bicycle and enter the cycle race this afternoon – you might do better than you've done with your sheep.'

'Cheeky sod!' I said. 'I'll get that hundred – without any help from you.'

Deep down, though, I was not so sure.

'What's your next show?' he asked.

'Moreton-in-the-Marsh.'

'You won't do any good down there. I would never

enter. The local breeds tend to do well there . . . breeds like the Oxford Down.'

'We'll see about that,' I said, confidently.

He was right, though. The Oxford Downs did beat us. They were mixed-breed classes, with a very large entry. We could manage only three prizes, and none higher than a third.

We had now won ninety-seven prizes at sixteen shows, and there was only one show left – Kington in Herefordshire. It had Kerry Hill classes, which meant that all my main rivals would be there. And they would all want to end up at the last show with a win!

We thought that our pair of ewes should win; after all, they had been unbeaten in Kerry Hill classes except for their very first outing at the Shropshire and West Midland Show.

But even if they did win, we were up against the fact that no prize is offered for the champion female Kerry sheep at Kington.

Could our rams, ram lamb and the other pair of ewes do the trick for us? It looked like being a monumental cliff-hanger. . . .

A week before the show I sold my car privately, and learnt that I would not be able to take delivery of my new vehicle for at least another fortnight. I needed a car to take the sheep to the show, and decided that I would hire one with a towbar for the day.

When I mentioned this to one of my farming friends, over a pint at the Black Horse, he said that I could borrow his car for nothing, provided I filled it up with petrol.

I could not miss a bargain like that. A couple of days before the show I went to have a look at the car, with the idea of getting used to it. It stood at the end of a farm drive . . . and it was a farm car, all right!

'It's taxed and insured,' said Bert proudly.

It was an estate car, of French origin. I had only seen anything to match it at a scrapyard.

'What do you call this?' I asked.

'It's my car!' said Bert.

'What colour is it?'

'Silver.'

'I would call it grey – and rusty brown,' I said. In fact, it was grey. Three-tone grey. The front end was darker than the rear end, and the passenger door was a different grey altogether. It was battered and dented, the sills were rusting away, and it looked unroadworthy, to put it mildly.

I peered through the dirty, cracked windows. Inside, it reminded me of Mick Hardy's old van. The floors were covered with hay, straw, soil, muck, sweet-papers. The seats were badly torn, and there was an assortment of bolts, spanners and aerosol cans in the glove compartment.

Bert opened the door – it dropped a couple of inches on its hinge – and showed me the controls.

'It's a complete wreck,' I said.

'Beggars can't be choosers,' he said resentfully.

It was the ideal sort of car for my poverty-pleading scrounging trips. When you are out to get parents to do jobs for the school farm, or to beg nails and wire and suchlike from farmers, it is fatal to turn up in a shiny new model. But now I needed one that would not let me down on the way to a show that had become the most important event in my life.

'I think I'll have to look elsewhere,' I said. At this late stage, though, I knew I would be lucky to find a car with a towbar and powerful enough to pull a trailer-load of sheep.

'This car will not let you down,' insisted Bert.

'I wouldn't like to bet on that.'

'It hasn't let me down yet, and she's fourteen years old – I've had her from new.'

'It's a pity you haven't looked after her, then,' I said.

I drove it home and parked it in the drive. I could see a couple of neighbours peering at it through their curtains. They seemed amused.

It took ages to prise off the petrol cap to fill the tank. When I went to check the oil and water, the bonnet would not budge. There must have been a way to do it, and I telephoned Bert to learn the secret. But he was out, and his wife could not help. In sheer desperation I attacked it with a tyre lever and a screwdriver, and at last I got it open. The oil and water were all right, but the bonnet was a devil to close again.

I met Vanessa and Matthew at the school gates the following morning. It was 5 o'clock and fairly dark; even so, they were not too complimentary.

'Do you honestly think that it's going to get us there, Sir?' asked Matthew.

'I've never seen such a wreck,' said Vanessa.

'The owner tells me that it has never let him down,' I said, trying to sound cheerful.

'I can't believe it's never let him down in fifty years,' said Matthew.

'Would you like me to walk in front with a red flag?' said a caustic Vanessa.

We hitched up the trailer. The lights would not work, but when we had loaded the sheep, it started to get lighter and I began to hope that we would not be stopped by the police for having no trailer lights.

It was extremely difficult to get into first gear, but we went smoothly along the M6 and M5. When I turned off the motorway, however, I found that first gear was now impossible to locate. It did not worry me much – I just started off in second gear.

After Leominster and its one-way system, we seemed to be on the wrong road.

'I'll navigate,' said Matthew.

'Get on with it,' I said. Matthew was an intelligent lad and I was sure he could put us on the right track. He told me to cut across country to save time, but after half an hour we seemed to be no nearer our destination.

I stopped the car and looked at the map. We were on the right road all right – but heading in the wrong direction.

I turned quickly, and saw a horse-box in front. 'We'll follow him,' I said. 'He's loaded up with horses, and there are rosettes in the windows – he's sure to be going to the show.'

The road became narrower and steeper, and the car started to splutter as we went up a single lane.

'He'll get us onto the main road,' I said.

A few seconds later the horse-box turned into a farm drive. I slowed down. The engine stalled.

'You've followed him back to his farm,' said Matthew.

I started the engine, but could not get into first gear . . . and the car would not pull the trailer in second on that hill. We began to roll backwards. I quickly applied the handbrake, then decided that I would have to try to reverse all the way down the hill. It would be tricky, but I am proud of my skill at reversing trailers. . . .

I reversed a little way, to discover that the bend – with a steep bank on one side and a drop on the other – was too sharp to negotiate. I was stuck!

'We'll have to unload the sheep,' said Vanessa.

'We'll lose them if we do that,' I said. 'They aren't going to stand in the road and wait for us to sort things out.'

'I think we can forget the show,' sighed Vanessa. 'I blame you, Matthew! Fancy not being able to do a bit of map-reading . . . and you're in the top geography group!'

'But I don't take geography as an option subject,' Matthew objected.

'Now he tells us!' screamed Vanessa.

'It was Mr Terry who suggested we followed the horse-box,' said Matthew sullenly.

'Stop panicking, Vanessa,' I ordered. 'Calm down.'

It was after nine-thirty, and judging would start promptly at ten. Even if we got out of this mess, we would be too late. How our rivals would love it!

'I suppose it's all my fault, really,' I groaned. 'If only I'd gone to the trouble of hiring a decent car!'

I took the rubber mats out of the car and pushed them under the rear wheels. Perhaps they would give us more grip. They would not. We needed mechanical help.

'Vanessa, Matthew – run up the farm drive and ask the farmer if he can help us out with a tractor and tow-rope,' I said.

They went, leaving me in gloomy silence. Funny . . . there was not even a bird singing. I waited, and looked at my watch, and studied the map to see where we had gone wrong. We seemed to be a twenty-minute journey away from the show. So near. . . .

Suddenly I heard the sound of a Land-Rover. The farmer drove out of his drive and backed towards me.

'You're in a right bloody mess, mate,' he said. 'Fancy trying to bring a load like this up this road.'

With that he secured a chain between the front of the car and the back of the Land-Rover, and pulled us smoothly over the brow of the hill, parking us on a stretch of flat road.

'That was easy,' he said with a big grin.

'Thanks very much,' I said. 'How much do I owe you?'

'Nothing.'

'I insist,' I said gratefully. 'Have a fiver.' My pupils could not believe their ears at this uncharacteristic benevolence.

I checked the way to the show. 'It's not far,' he said. 'Keep on going to the end of this lane – turn left, take the next right, and that'll get you onto the main road. Turn left and carry on for just a few more miles. That will take you into Kington.'

He added: 'I still can't work out how you managed to get here!'

'How long will it take me?'

'With a bit of luck, about ten minutes.'

'Are there any more steep hills, then?'

'No . . . it's all plain sailing.'

The steward let us into the showground at ten minutes past ten, and I continued to drive like a lunatic until I skidded to a halt at the sheep pens. Amazingly, the judging had not started.

'You took your time, John,' shouted Mr Williamson. 'Have you had a good journey?'

We unloaded the sheep into a couple of pens – there was no time to sort them out and pen them properly. The judge was waiting and the other exhibitors were already leading out the two-year-old rams. I threw on my white coat, quickly put a halter on William, and began to lead him into the ring.

A steward bawled. 'Move that car and trailer immediately – it's in the way!'

'I'll move it straight after the judging,' I said politely. 'I'm too busy at the moment.'

Some of these stewards get power-mad when they are given a badge! He insisted: 'If you don't move it I'll report you to the secretary.'

I did not know whether to lose my temper, or to ignore him. I pretended not to hear, and he disappeared. I got on with the job in hand – showing William.

He was placed fourth in his class.

'That's ninety-eight prizes!' shouted Vanessa.

It was the shearling ram class next. Our Benji came second.

'Ninety-nine!' shrieked Vanessa. She and Matthew linked hands and did a kind of war-dance.

Vanessa and I got the ewes out in the ring early. They stood like marble statues – and my heart was thumping like a sculptor's hammer.

The judge seemed to like them. After an age, he told us to stand our ewes at the top of the line, and told Mr Williamson to put his ewes next to ours, in second place.

But he had not finished comparing the two pairs of ewes.

'I'm going to beat you today,' said Mr Williamson confidently.

'Not if we can help it,' I replied.

The judge then placed ours first and his second. Mr Porter's ewes were third.

We had got our hundred.

Our other pair of ewes, being shown by Matthew with the help of a friendly breeder, took the fourth prize. In the ewe lamb class that followed we got a further fourth.

We had won a hundred and two prizes in the year.

I felt as though an enormous burden had disappeared. I had achieved my goal and won my bet. I could relax again now. . . . Well, almost. We still had to get home!

We did that without any more problems, and in a merry mood. On the way I remember Matthew saying, with a happy sigh: 'There must be no better way of spending summer than loading sheep on and off trailers.'

I warmly agreed with that. We had travelled almost three and a half thousand miles to seventeen shows. Our ewes had achieved wonders. They had won fourteen first prizes, they had been three times supreme Kerry Hill exhibit, and they were six times female Kerry Hill champions.

In its next issue the *Farmers Guardian* summed it all up with the headline: 'Terry's Kerrys achieve ton-up!'

I felt that our magnificent obsession had been well worth while and one of the best parts was collecting Mr Williamson's fiver.

Surprise, Surprise . . .

EVERY TIME I walked across the playground to my department, I noticed the fencing. I suppose that to most people it looked quite presentable – but to my eyes it spoilt the department and badly needed replacing.

We originally fenced in our three grass paddocks in December 1977, using Scots pine posts and rails. Netting was stapled along the back of the fence to keep lambs out of the playground.

The posts and rails had been creosoted about fifteen times. Even so, some of the posts were rotten, and we had done a fair amount of renovating.

In many places the posts were no longer upright. They leant away from the playground, for two main reasons. Firstly, pupils playing football had been climbing over them for years to retrieve their footballs. Secondly, they were a favourite trysting prop for young courting couples.

Picture the scene . . . the girl has her back against the fence. The lad faces her, enfolding her in his arms. They kiss, oblivious to the creaky protest of the fence. It is not built to withstand their pleasurable pressures.

One of the fifth-year girls, Jackie, was often enmeshed in this posture by her ardent swain, Brian. It was a case of passion in the playground – sometimes even before school started in the morning.

Brian was very popular with the girls, a real 'Jack-the-lad' type who always seemed flush with money. He wore expensive clothes and a chunky medallion around his neck.

I often warned the pair to stop leaning on my fence, suggesting that they might be more comfortable elsewhere. Once, I even toyed with the idea of throwing a bucket of water over them.

One morning, Mr Beech brought a visitor to look at the school farm. Mr Perks, a headmaster from Leicestershire, was keen to set up a Rural Studies department at his school. I think he was impressed with our set-up and he asked many questions about the syllabus we followed, how I catered for the less-academic pupils, our average class size, and so on.

My guided tour ended at the start of the morning break, and Mr Beech was keen to take our guest to coffee and biscuits. There were pupils all over the playground, playing football, chatting, having a giggle. And, of course, Jackie and Brian were leaning against their favourite fence-post.

I noticed that Mr Beech was watching them out of the corner of his eye, and that he was not at all pleased.

I shook hands with Mr Perks, and was about to say cheerio when he said: 'Just before I go, Mr Terry . . . tell me, how do you segregate the sexes?'

'With a bloody crowbar!' I said, pointing at Jackie and Brian. This spontaneous comment earned me one of Mr Beech's blackest scowls. . . .

I moved the loving couple on several occasions after that . . . until the morning when one of my fourth-year groups made an excellent job of creosoting the fence, completing their task with a large notice: 'Wet creosote'.

At afternoon break Jackie and Brian burst into my classroom. She was in tears and he was in a raging temper. When Jackie turned round I could see the problem – two parallel brown stripes running across the back of her blouse. Brian's hands were also brown, and so was one of the cuffs on his expensive sheepskin coat.

'What are you going to do about this?' demanded Brian.

I countered with a deep sigh. 'How many times,' I asked, 'have I told you not to lean on the fence – and to do your courting somewhere else, preferably out of school hours?'

'I don't know, Sir.'

'Well . . . once, twice, ten times?'

'About ten, Sir,' he moaned.

172

'You should have known better,' I asserted.

'My mum will go mad when she sees my blouse,' sobbed Jackie. She added: 'I think we should send you the cleaning bill.'

'You can send it, but I won't pay it,' I said. 'Follow me.'

They followed at my heels like a couple of terrier dogs in search of a titbit. I pointed at the sign on the fence.

'You both need a pair of glasses!' I snapped.

'Yes, Sir,' they said in unison. They walked away without another word. I never saw them leaning against the fence again.

It had been the best fence we could afford when it was erected, but the posts had not been set in concrete. It had given fairly good service, and it had now passed its best. I decided on a smart replacement – one that would really last.

I opted for mortised post-and-rail construction, with the mortised posts nine feet apart, four rails, and a strengthening post – known as a prick-post – in the centre of each bay.

It made an excellent class exercise to measure up the paddocks; a good lesson in practical mathematics. Each pupil was asked to work out how many posts and rails would be needed, and it was interesting to note how many pupils remembered to add an extra post for the end of every run of fencing.

The exercise was carried out with three different groups, and when they all agreed with my figures I ordered the materials. The 6 ft 6 in posts, in 3-inch square timber, were mortised four times. The rails were 3½ inches by 1½ inches. We had six runs to do – a total of 133 yards.

I enlisted the help of a parent and some of my strongest fifth-year pupils to do the job. We dug out the first hole, put in the post, and pushed the rails in through the mortice. The other ends of the rails showed us where to dig the next hole. The prick-post was placed in the middle at the back, and the rails were nailed to it.

Farmers would probably put this smaller post on the side of the fence nearest the road. We reversed this, because our pushing comes mainly from the playground and road side rather than from the stock in the paddock.

Fencing is a rewarding job; you can certainly see where you have been! A chainsaw made it easier going than our old rip-saw – and stopped our neighbours from dozing in the afternoons. After making sure that all the posts were vertical in their holding-beds of soil, we put concrete around them.

There was no need to creosote the new fence, as it was all tanalised timber. We stapled wire sheep-netting at the back of it. Where the paddocks were adjacent to back gardens, there was no need for the mortised fencing, and we drove in 5 ft 6 in tanalised stakes and attached sheep netting surmounted by two lines of high-tensile wire.

The finished fencing looked superb, and two coats of white paint on the gates put the finishing touch to the job. With our Kerry Hill sheep grazing in the paddocks, it presented a splendid sight to anyone coming up the school drive. We sorted the old posts and rails and sold the sound ones to help to pay for the new fence.

The new-look paddock gates prompted us to turn to the rest of the paintwork on the school farm. The original blue we had used on all the doors of the buildings was no longer in manufacture, and we decided to choose a slightly darker Oxford blue. New stable-type doors replaced the rotten ones on the pig building and feedstore, and were painted with primer, undercoat and topcoat.

Other doors on the brick building and the large wooden building needed just a topcoat. Counting a split stable door as two doors, we had a total of eighteen doors to paint, back and front, plus six window frames inside and out. To make it all look really professional, we painted all door hinges and bolts black.

Our large wooden building is creosoted each autumn, ready for winter's ravages. Our brick building is rendered outside, and needs no painting there; the plastered inside walls we redecorated with white emulsion, moving the livestock around to give access to each section. Some of my pupils are excellent painters; others seem to put more paint on the floor than on the object they are supposed to be treating!

They are not the only messers, admittedly. Within minutes of putting our four new pigs into their newly

painted building, one made a large brown blot on the gleaming wall.

'I can't believe it, Sir . . . after all the work we've done,' said a dejected Samantha. . . .

We now prepared for another big job which had to be done before winter set in. The snowberry hedge parallel to our large wooden building looked very untidy – although we had cut it regularly, it was thin in some places and taking over the farm in others. I was keen to get it dug out and replaced with heathers and dwarf conifers.

My fifth-year pupils were keen to take the job on – it would be a good project for them, and some of the younger pupils would help with the donkey work.

To use a mechanical digger would make a mess of our lawn and rosebeds. We had to do it by hand. We cut the hedge down almost to ground level for a start, then cleared away all the top growth. That was easy – the hard part would be the removal of the long row of stumps that remained.

The hedge was sixty feet long and we had ten spades. Ten stout volunteers started to dig, pausing only when they found that they were being watched by an interesting – not to say notorious – visitor.

It was Clive Gibson, one of my former pupils who was 'done' for grievous bodily harm while still at school.

'How are you, Clive?' I asked.

'I'm all right, Mr Terry,' he said. 'But I'm going through an identity crisis.'

'Oh . . . you're trying to find your true self . . . what life is all about?'

'No, Mr Terry . . . I'm worried that a security guard might be able to identify me!'

He was joking, I hoped. . . .

'I've come to see if I can borrow your lawnmower,' he went on, with an all-too-innocent grin.

It was my turn with the quick repartee: 'Of course you can,' I said, 'as long as you don't take it out of our garden.' He was persuaded to stay and help dig out a few roots, and made very light work of it.

Next day, rain put a stop to our project. The following day, also, was really too wet, and we packed up after half an hour. The sludgy wellingtons of the lads made a dreadful mess of the paths, which had to be swilled down.

'Look at all this mess from your wellingtons,' I stormed.

'I've got three wellingtons, Sir,' shouted one of the offenders. 'I've grown another foot since I've been at this school!' That one caused more groans than the weather. . . .

All through the following week we worked on the roots, using crowbars for leverage – even so, one spade handle was sundered – and when most of them were out we levelled off the ground and dug it over, removing more bits of root. After leaving it to settle we dug it once more – I was keen to get every fragment of root removed to avoid them shooting up again.

Our rectangular border now looked very neat. The shape was too formal, however. One long side and the two short sides were bordered by paved paths, and the other long side adjoined a lawn. This side could easily be altered to produce a less formal shape.

I drew some curves to scale on graph-paper and invited each fifth-year pupil to create an alternative plan. My plan was not the most pleasing, and most of us voted for the one drawn up by Helen – an 'arty' girl.

I had contacted a specialist nursery to supply us with pot-grown heathers, dwarf conifers and peat – at a good discount, of course – and our next exercise was to work out exactly what we needed.

Heathers are planted at 30–45 centimetres between each plant, and we decided to plant contrasting groups of them, interspersed with a few specimen conifers. Our aim was for flowers and foliages which would provide a contrast in colour throughout the year – therefore we did not want all the summer-flowering heathers at one end of the border and the winter flowerers at the other.

We settled for fifteen heathers in each group, making two hundred and fifty altogether – plus fourteen dwarf conifers and a blue cedar tree. Four bales of peat completed the order. Heathers like an acid soil, but ours was neutral – so the peat was to increase the acidity.

The heathers arrived early one morning, all labelled, and we planted them that afternoon. First we spread peat all over the border to a depth of about five centimetres; then we used a stick to mark out the planting areas in accordance with our scale plan. We watered the heathers and placed them into position, on top of the soil still in their pots. It all looked fine, and so we carried on, took them out of their pots and got them planted, putting more peat into the bottom of each hole and watering each plant well.

We had not finished by the end of the last lesson, but I carried on with a team of keen pupils and finished the job after school. Finally we got the hosepipe out and soaked the whole border, and stood back, tired and happy, to admire the remarkable transformation we had achieved in a single day.

Winter was drawing on and I mentally took stock of the goals we had attained in an outstanding year.

We had continued to breed and show our sheep, successfully – culminating in our second championship triumph at the Royal Show. Our department presented a very different picture from that of 1974, when I began to develop it with scrounged and secondhand materials. We now had a new trailer, permanent paving and new fencing. Our buildings – and our heather garden – were joys to behold.

Our flock of Kerry Hill sheep had grown to twenty breeding ewes, and our other livestock – calves, goats, pigs, rabbits, laying hens, ducks and geese – were all doing fine. We had a well-stocked greenhouse and excellent flower, fruit and vegetable gardens.

Our pupils were now able to become proficient in a wide range of farm, garden and experimental tasks. After leaving us, some went on to careers in agriculture

or horticulture having taken college courses in these disciplines. David Harper, for example, now had a good job as full-time cowman on Lord Clifton's estate, after completing his studies at the local college of agriculture. Other pupils had jobs connected with agricultural businesses.

Many more had not forgotten what I had taught them; some had married and set up a home where they were growing vegetables, flowers and fruit and keeping small livestock.

'What is your next ambition, Sir?' asked Elaine one day.

'You can always find something more to do . . . new challenges . . . something better to aim for,' I replied.

'Such as . . .?'

'We could do with some more land nearby, so we could keep more livestock.'

'Mr Beech hasn't given you permission to use the cricket, football or hockey pitches, has he?' queried Gail.

'Not yet. . . .'

'How about digging up some of the playground?' Always the joker, Matthew.

'I can't see Mr Beech or Warwickshire County Council looking too kindly on that.'

Vanessa probed: 'What other ambitions do you have in life, Mr Terry?'

I mused for a moment. 'Well . . . I live on a modern housing estate. What I would really like, though, is a nice house in the country, with some land, so I could wake up in the morning and see Kerry Hill sheep grazing when I looked out of the bedroom window. That would be a lovely sight! But I'll have to win the pools before that happens . . . and I don't do the pools, so it's only a dream. . . .'

There was not much room for dreaming in the department. My life was mainly made up of feeding and looking after livestock in the early morning, registration, assembly, lessons both practical and theoretical, various other school duties, and feeding and looking after livestock at evenings and weekends.

There were endless exercise books and projects to mark, examinations to set – a thousand and one other jobs to do. . . .

On my birthday I awoke and lay in the half-light, thinking that I ought to make an effort and get out of bed, when I thought I heard a sheep baaing.

I started to doze – and, yes, there it was again.

I began to wonder whether I had dropped off in the sheep pen at school. But it *was* my bedroom – and that was most definitely another baa I was hearing from outside.

I decided to get up and investigate. Most likely there was a lorry or trailer parked outside with sheep in it – perhaps one of my farmer friends was calling to see me on his way to market, and would be knocking the door at any moment.

I moved the curtains and peered out of the window. My mouth gaped in astonishment and I rubbed my eyes to make sure I was not seeing things.

The front lawn was enclosed with electric sheep netting, and grazing there – right under my window – were four Kerry Hill ewes.

I could not move. . . . I could do nothing but stare.

There was no one else in sight. What an extraordinary

181

scene – a modern bungalow on an urban housing estate with four sheep in the front garden!

And they were our sheep, I could recognise them anywhere. Tessa, Amanda, Ann and Lorna.

I put my clothes on and ran out into the drive, to find the postman scratching his head as he looked at them in amazement.

I was trying to figure out what had happened. Could it be a move by some of the school's neighbours whose back gardens were next to the sheep paddocks? Perhaps they wanted to see how I would like noisy sheep close to my living-quarters. . . .

Surprise, Surprise . . .

More than likely it was one of those television programmes specialising in practical jokes. We had been studied by a lot of film crews. Maybe one of them had put forward our quaint little farm as an ideal subject for a 'candid' caper. . . .

The sheep were making the most of the grass on the lawn – and they had already started to fertilise it. I saw curtains moving in the upstairs of a house across the road, but apart from the postman there was no one about.

I was wondering what to do next when five heads suddenly bobbed up from behind a car parked on the other side of the road.

'Surprise . . . surprise. . . . Happy birthday, Sir!' they chanted.

It was five of my sheep 'specialists' – Vanessa, Elaine, Gail, James and Matthew.

'What the . . .' I began, but I could not finish – I was silenced by the proverbial lump in the throat. . . .

They all ran up, laughing and singing their greeting.

'You did say, Sir,' puffed Vanessa, 'that your ambition was to wake up in the morning and see Kerry Hill sheep grazing by your bedroom window!'

'Did I? I . . . don't really remember. . . .'

'Well, you did, Sir.'

'How did you get them here?'

'We walked them from school on their halters,' said Elaine proudly. 'They've all been shown, so it was easy!'

'Well . . . you certainly caught me out,' I said. 'It's given me quite a shock.'

By this time a small crowd had gathered at the scene. The postman was joined by the paper boy, the milkman, and a dozen neighbours. The sheep were enjoying the attention – two small children were already giving them some bread.

'You don't *mind*, do you, Sir?' asked Vanessa. 'We were a bit worried. We didn't know how you'd react.'

'No!' I replied. 'I don't mind.... Thank you for a wonderful birthday present!'

Also published by Farming Press

The cartoons in *Pigs in the Playground* and *Calves in the Classroom* were drawn by Henry Brewis who lives on a cereal and livestock farm in Northumberland.

Farming Press have published three of his books:

Funnywayt' mekalivin' and *The Magic Peasant* are both agricultural stews of cartoons and verses featuring Sep, the universal peasant. His world includes collie dogs, auctioneers, cows, ewes, the long-suffering wife and a range of farm visitors, welcome and unwelcome.

Don't Laugh Till He's Out of Sight is a collection of the best of Henry Brewis's writing, revealing the hazards awaiting anyone venturing on life as a farmer.

Farming Press publish some 50 books on agriculture and animal health. Among these is a small range designed for young people embarking on their agricultural education at school or college. Specifically the level is City and Guilds I and II.

Farm Crops, Graham Boatfield
Farm Livestock, Graham Boatfield
Farm Machinery, Brian Bell
Farm Workshop, Brian Bell
Calculations for Agriculture and Horticulture,
    Graham Boatfield and Ian Hamilton

If you would like the details of any of these books or a free illustrated catalogue please contact:

Books Department 3, Farming Press Limited,
Wharfedale Road, Ipswich IP1 4LG.